Contents

How To Use This Book

The ***Year 6 Targeting HASS Activity Book*** is filled with exciting stimulus materials to assist you in developing a greater understanding of the world you live in—its past and its present. The activities have been designed to encourage you to share opinions, think creatively, analyse ideas and information and respond in a variety of ways.

Each unit delivers content from the knowledge and understanding strand of the Australian Curriculum and supports the key inquiry questions in each subject area. The units also allow you to develop inquiry skills through:

Researching

Students identify and collect information, evidence and data from primary and secondary sources.

Questioning

Students develop questions about events, people, places, ideas, developments and issues to guide their investigations and satisfy their curiosity.

Analysing

Students explore information, evidence and data to identify and interpret features, patterns, trends and relationships, key points, facts and opinions and points of view.

Communicating

Students present ideas, findings, viewpoints, judgements and conclusions in digital and non-digital forms for different audiences and purposes.

Evaluating and Reflecting

Students propose explanations for events, developments and issues, draw evidence-based conclusions and use criteria to make informed decisions and judgements.

This book is organised into the Year 6 curriculum areas, as follows:

- **Units 1–12:** **HISTORY**
- **Units 13–24:** **GEOGRAPHY**
- **Units 25–28:** **CIVICS AND CITIZENSHIP**
- **Units 29–32:** **ECONOMICS AND BUSINESS**

TARGETING HASS 6 © PASCAL PRESS ISBN: 9781925726077

The ***Targeting HASS Activity Book*** series is designed to be a flexible learning resource. The units do not have to be completed in numerical order. Your teacher may direct you to complete units and activities from one learning area before moving on to a different learning focus. In this way, the *Targeting HASS* series will complement any school's scope and sequence.

The seven **Australian Curriculum's General Capabilities** listed below are embedded in the units of this book:

- Literacy
- Numeracy
- Information and Communication Technology (ICT) Capability
- Critical and Creative Thinking
- Personal and Social Capability
- Ethical Understanding
- Intercultural Understanding

A number of the activities and questions are open-ended. This allows for student challenge and differentiation. The questions are written to encourage you to think deeply about topics and provide extended responses. For some open-ended questions, answers are not provided. These questions are best graded through peers marking each other's work, or through student–teacher conferences. There are brief sample responses to guide the assessment of your work for some open-ended questions.

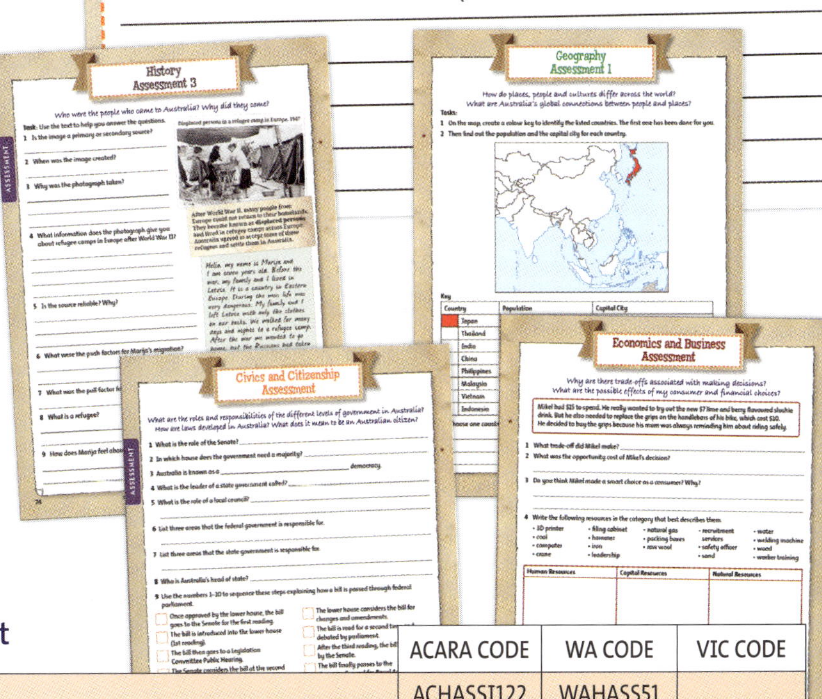

The final section features eight **assessment** activities:

- 3 History
- 3 Geography
- 1 Civics & Citizenship
- 1 Economics & Business.

Each assessment activity relates to a key inquiry question. Make sure you have completed the units for that inquiry question before tackling the assessment.

Additional inquiry skills covered by each unit

		ACARA CODE	WA CODE	VIC CODE
Questioning	Develop questions to guide an inquiry	ACHASSI122	WAHASS51	
Researching	Locate and collect information and data from primary and secondary sources	ACHASSI123	WAHASS52	VCGGC088
	Organise and represent data in a range of formats	ACHASSI124	WAHASS58	VCGGC088
	Sequence information	ACHASSI125	WAHASS56	VCHHC082
Analysing	Examine primary and secondary sources to determine origin and purpose	ACHASSI126	WAHASS55	VCHHC083
	Examine different viewpoints	ACHASSI127	WAHASS57	VCHHC084
	Interpret data and information in a range of formats	ACHASSI128	WAHASS56	VCGGC090
Communicating	Present ideas, findings and conclusions	ACHASSI133	WAHASS61	
Evaluating and Reflecting	Evaluate evidence to draw conclusions	ACHASSI129	WAHASS59	
	Use criteria to make decisions and judgements	ACHASSI131	WAHASS60	
	Reflect on learning to propose action and predict the probable effects	ACHASSI132	WAHASS63	

Curriculum correlations

	History											
	1	2	3	4	5	6	7	8	9	10	11	12
ACARA CODE: DESCRIPTION – HISTORY												
ACHASSK134 Key figures, events and ideas that led to Australia's Federation and Constitution	✔	✔	✔	✔								
ACHASSK135 Experiences of Australian democracy and citizenship, including the status and rights of Aboriginal and Torres Strait Islander Peoples, migrants, women and children					✔	✔	✔	✔				
ACHASSK136 Stories of groups of people who migrated to Australia since Federation (including from ONE country of the Asia region) and reasons they migrated								✔	✔	✔		
ACHASSK137 The contribution of individuals and groups to the development of Australian society since Federation											✔	✔
ACARA CODE: DESCRIPTION – GEOGRAPHY												
ACHASSK138 The geographical diversity of the Asia region and the location of its major countries in relation to Australia												
ACHASSK139 Differences in the economic, demographic and social characteristics of countries across the world												
ACHASSK140 The world's cultural diversity, including that of its indigenous peoples												
ACHASSK141 Australia's connections with other countries and how these change people and places												
ACARA CODE: DESCRIPTION – CIVICS AND CITIZENSHIP												
ACHASSK143 The key institutions of Australia's democratic system of government and how it is based on the Westminster system												
ACHASSK144 The roles and responsibilities of Australia's three levels of government												
ACHASSK146 Where ideas for new laws can come from and how they become law												
ACHASSK147 The shared values of Australian citizenship and the formal rights and responsibilities of Australian citizens												
ACHASSK148 The obligations citizens may consider they have beyond their own national borders as active and informed global citizens												
ACARA CODE: DESCRIPTION – ECONOMICS AND BUSINESS												
ACHASSK149 How the concept of opportunity cost involves choices about the alternative use of resources and the need to consider trade-offs												
ACHASSK150 The effect that consumer and financial decisions can have on the individual, the broader community and the environment												
ACHASSK151 The reasons businesses exist and the different ways they provide goods and services												
NSW HISTORY STAGE 3 CURRICULUM OUTCOMES												
HT3-1 Describes and explains the significance of people, groups, places and events to the development of Australia											✔	✔
HT3-3 Identifies change and continuity and describes the causes and effects of change on Australian society	✔	✔	✔	✔	✔	✔	✔	✔	✔	✔		
HT3-4 Describes and explains the struggles for rights and freedoms in Australia, including Aboriginal and Torres Strait Islander peoples					✔	✔	✔	✔				
HT3-5 Applies a variety of skills of historical inquiry and communication	✔	✔	✔	✔	✔	✔	✔	✔	✔	✔	✔	✔
NSW GEOGRAPHY STAGE 3 CURRICULUM OUTCOMES												
GE3-1 Describes the diverse features and characteristics of places and environments												
GE3-2 Explains interactions and connections between people, places and environments												
GE3-3 Compares and contrasts influences on the management of places and environments												
GE3-4 Acquires, processes and communicates geographical information using geographical tools for inquiry												

Geography												Civics & Citizenship				Economics & Business				History Assessment			Geography Assessment			Civics & Citizenship Assessment	Economics & Business Assessment
13	14	15	16	17	18	19	20	21	22	23	24	25	26	27	28	29	30	31	32	1	2	3	1	2	3		
																				✓							
																					✓						
																						✓					
✓	✓	✓																					✓	✓	✓		
	✓		✓			✓					✓														✓		
				✓					✓		✓													✓			
					✓	✓	✓	✓		✓															✓		
													✓														
																										✓	
														✓												✓	
															✓											✓	
												✓															
																✓											✓
																		✓	✓								✓
																	✓										
																				✓	✓	✓					
																					✓						
																					✓	✓					
✓	✓	✓	✓	✓	✓	✓	✓	✓	✓	✓	✓												✓	✓	✓		
				✓	✓	✓	✓	✓	✓	✓	✓													✓			
✓	✓	✓		✓					✓		✓												✓	✓	✓		
✓	✓	✓	✓	✓		✓	✓	✓	✓	✓	✓												✓				

Curriculum correlations

	History											
	1	2	3	4	5	6	7	8	9	10	11	12
VICTORIA HISTORY LEVELS 5 & 6 CURRICULUM OUTCOMES												
VCHHK093 The significance of key figures and events that led to Australia's Federation, including British and American influences on Australia's system of law and government	✔	✔	✔	✔								
VCHHK094 The different experiences and perspectives of Australian democracy and citizenship, including the status and rights of Aboriginal and Torres Strait Islander peoples, migrants, women, and children					✔	✔	✔	✔				
VCHHK095 The stories and perspectives of people who migrated to Australia, including from one Asian country, and the reasons they migrated								✔	✔	✔		
VCHHK096 Significant contributions of individuals and groups, including Aboriginal and Torres Strait Islander peoples and migrants, to changing Australian society											✔	✔
VICTORIA GEOGRAPHY LEVELS 5 & 6 CURRICULUM OUTCOMES												
VCGGK090 Interpret maps and other geographical data and information using digital and spatial technologies as appropriate, to develop identifications, descriptions, explanations and conclusions that use geographical terminology												
VCGGK092 Location of the major countries of the Asian region in relation to Australia and the geographical diversity within the region												
VCGGK093 Differences in the demographic, economic, social and cultural characteristics of countries across the world												
VCGGK096 Environmental and human influences on the location and characteristics of places and the management of spaces within them												
VCGGK097 Factors that influence people's awareness and opinion of places												
VICTORIA CIVICS & CITIZENSHIP LEVELS 5 & 6 CURRICULUM OUTCOMES												
VCCCG008 Discuss the values, principles and institutions that underpin Australia's democratic forms of government and explain how this system is influenced by the Westminster system												
VCCCG009 Describe the roles and responsibilities of the three levels of government, including shared roles and responsibilities within Australia's federal system												
VCCCG010 Identify and discuss the key features of the Australian electoral process												
VCCCG011 Identify the roles and responsibilities of electors and representatives in Australia's democracy												
VCCCL012 Explain how state/territory and federal laws are initiated and passed through parliament												
VCCCL013 Explain how and why laws are enforced and describe the roles and responsibilities of key personnel in law enforcement, and in the legal system												
VCCCC014 Identify who can be an Australian citizen and describe the rights, responsibilities and shared values of Australian citizenship and explore ways citizens can participate in society												
VCCCC015 Identify different points of view on a contemporary issue relating to democracy and citizenship												
VCCCC017 Examine the concept of global citizenship												
VICTORIA ECONOMICS & BUSINESS LEVELS 5 & 6 CURRICULUM OUTCOMES												
VCEBR002 Explore the concept of opportunity cost and explain how it involves choices about the alternative use of limited resources and the need to consider trade-offs												
VCEBR003 Identify types of resources (natural, human, capital) and explore the ways societies use them in order to satisfy the needs and wants of present and future generations												
VCEBC005 Consider the effect that the consumer and financial decisions of individuals may have on themselves, their family, the broader community and the natural, economic and business environment												
VCEBE010 Make decisions, identify appropriate actions by considering the advantages and disadvantages, and form conclusions concerning an economics or business issue or event												
VCEBB006 Identify the reasons businesses exist and investigate the different ways they produce and distribute goods and services												

TARGETING HASS 6 © PASCAL PRESS ISBN: 9781925726077

Geography												Civics & Citizenship				Economics & Business				History Assessment			Geography Assessment			Civics & Citizenship Assessment	Economics & Business Assessment
13	14	15	16	17	18	19	20	21	22	23	24	25	26	27	28	29	30	31	32	1	2	3	1	2	3		
																				✔							
																					✔						
																						✔					
✔	✔	✔																					✔	✔	✔		
✔	✔	✔																					✔	✔	✔		
✔	✔	✔																					✔	✔	✔		
			✔	✔		✔			✔		✔																
					✔	✔	✔	✔		✔																	
													✔														
													✔														
													✔														
													✔														
														✔													
														✔													
															✔												
												✔			✔												
												✔															
																✔											
																✔	✔	✔									
																		✔	✔								
																✔			✔								
																	✔										

To Federate or Not to Federate?

Before 1901, Australia was not a nation. Instead, it was made up of six separate colonies which Britain still largely controlled. Each colony acted like an individual country. They had their own governments and rules, as well as their own armies, rail gauge (distance between tracks), stamps and taxes. These differences caused problems, especially when moving people and goods from one colony to another.

Some people thought it would be a good idea if the colonies federated (joined together) and became one nation. This new nation could have one federal government that would make laws about issues that affected the whole country, such as defence and transport. Each colony would become a state and could also have their own state governments, which could control state issues. Some people liked the idea and some did not. In the end, the majority voted for Federation. On 1 January 1901, the colonies federated and became the Commonwealth of Australia.

THE ARGUS, WEDNESDAY, JUNE 1, 1898.

THE GREAT WORK FOR FRIDAY.

BARRIERS BETWEEN BROTHERS.

SHALL THEY REMAIN?

"LOOK UPON THIS PICTURE.

AUSTRALIA

AND ON THIS."

Cartoon from 1898 newspaper urging the colonies to federate

Reasons for Federation

- ✔ Federation would remove taxes and allow free trade. Goods were taxed as they crossed each colony's borders — this made them expensive.
- ✔ There would be one army, one postal service and the same rail gauge across the country.
- ✔ The nation could share natural resources, such as water from rivers that ran through different colonies.
- ✔ Australians wanted to make their own rules about issues that affected them, instead of the British making them.
- ✔ The majority of the population had been born in Australia and wanted to be recognised as Australian, not British.
- ✔ Many thought the new federal government would create laws to ensure that only white people of mainly British heritage could live in Australia. At the time, this was considered a good thing.

Reasons against Federation

- ✗ Rich colonies did not want to share their wealth with poorer colonies.
- ✗ Poorer colonies liked the tax system. Taxing goods that crossed their borders was one of the few ways they made money.
- ✗ Poorer or more remote colonies thought the rich colonies would not listen to them and would take control.
- ✗ It was too hard. Too many important decisions would need to be made, such as where the capital would be and what form of government Australia should use.

Source: Australian Geography Centres: *Upper Primary*, p. 25, Blake Education

TARGETING HASS 6 © PASCAL PRESS ISBN: 9781925726077

Research

Find another cartoon from the 1890s which looked at the issue of Federation.

What does it tell you about people's opinions and reasons for wanting or not wanting change?

A good place to start is the State Library of South Australia: https://federation.collections.slsa.sa.gov.au/images_index.htm

____

____

____

____

____

____

____

____

Analysing

Is the cartoon used in the text a primary or secondary source?

____

Is the cartoon in favour or against Federation?

____

How does the cartoon convey its point of view?

____

____

____

____

Questioning

Imagine you were a reporter interviewing a state politician about whether or not Australia should Federate. Write four questions you would ask.

1 ____

____

____

2 ____

____

____

3 ____

____

____

4 ____

____

____

Communicating

Create a PMI chart to show the Positive, Minus and Interesting aspects of Australia becoming a Federated nation.

Share and discuss your chart in a small group.

Evaluating and Reflecting

Has Federation been good for Australia? Give reasons for your response.

____

____

____

____

____

UNIT 2

Timeline Towards Federation

1889 Tenterfield Oration
Henry Parkes gives a famous speech at Tenterfield. He calls for the Australian colonies to become one nation with a federal government.

1890 Australasian Federation Conference
The delegation decides that Australia will follow the British parliamentary system and use the United States Constitution as an example.

1891 First Constitutional Convention
A convention is held in Sydney over five weeks. Politicians from all Australian colonies and New Zealand attend. Many discussions are held on how to federate. The first draft Constitution is written and approved.

Members of the Australasian Federation Convention, 1891

1893 The Corowa Plan
During a meeting in Corowa, John Quick proposes that a new draft Constitution be written. He suggests that it be written by elected state representatives and that the public should vote on the draft.

1897–98 Second, Third and Fourth Constitutional Conventions
Three more conventions are held. The delegates first decide not to use the 1891 draft Constitution. Between conventions, the delegates return to their colonies to discuss the new drafts and to make changes. The changes are discussed at the following conventions until a final draft is decided upon.

1898 First Referendum held in NSW, Vic., SA and Tas.
The people vote on whether to accept the draft Constitution or not. The 'yes' vote wins in all colonies that participate, except New South Wales.

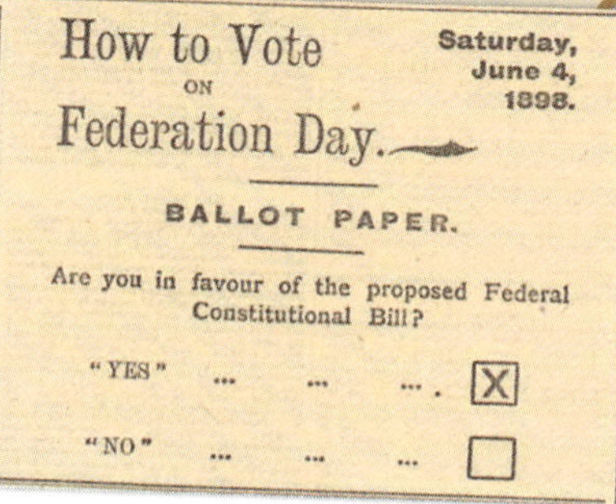
How to Vote on Federation Day.

Saturday, June 4, 1898.

BALLOT PAPER.

Are you in favour of the proposed Federal Constitutional Bill?

"YES" ... ... [X]

"NO" ... ... [ ]

1899 Second Referendum held in NSW, Vic., SA, Tas. and Qld
Changes are made to the Constitution and the public votes again. The 'yes' vote wins in all colonies that participate.

1900 *Commonwealth of Australia Constitution Act* passed by British Parliament
A group of politicians travels to London to present the Constitution to the British Government. The *Commonwealth of Australia Constitution Act* is passed so Australia can federate.

1900 Western Australia joins the Federation Movement
Western Australians vote in a third referendum on whether or not to federate. The 'yes' vote wins by a landslide.

1901 Federation
On 1 January 1901, all of the colonies of Australia federate (join together) and become one nation known as the Commonwealth of Australia.

Source: Australian History Centres: *Upper Primary*, p. 41, Blake Education; *Blake's Australian History Guide*, p. 68–69, Pascal Press

UNIT 2

Research

Discover five interesting facts about the Tenterfield oration.

1 ______________________________

____

2 ______________________________

____

3 ______________________________

____

4 ______________________________

____

5 ______________________________

____

Questioning

Write four questions to check a friend's understanding of the text.

1 ______________________________

____

____

2 ______________________________

____

____

3 ______________________________

____

____

4 ______________________________

____

____

Analysing

Which event on the timeline do you think was the most significant in helping to achieve the 'Yes' vote in favour of Federation? Why?

____

____

____

____

____

____

____

Communicating

To celebrate Australia's federation, a commemorative mug was produced. It featured an image of Sir Henry Parkes.

Design a mug to commemorate the next Federation Day. Sketch what it would look like. In small groups, discuss your choices of patterns and images.

Evaluating and Reflecting

Should Australia celebrate Federation Day on 1 January in the same way that we celebrate Australia Day on 26 January? Explain your point of view.

____

____

____

____

____

UNIT 3

HISTORY

Important Federation Australians

1889
Sir Henry Parkes
(1815–1896)

Tenterfield Oration

Henry Parkes had emigrated from England in 1839. He worked as a shopkeeper and journalist before entering politics. He was elected Premier of New South Wales five times. He is often called 'the father of Federation' because of a famous speech he made at Tenterfield, New South Wales, in 1889. This speech became known as the Tenterfield Oration. In it, Parkes proposed that the Australian colonies join together to form one nation. He challenged the audience:

"... whether the time has not now arisen for the creation on this Australian continent of an Australian Government, as distinct from the local governments, and an Australian Parliament."

Parkes said the United States of America and Canada had achieved what he wanted Australia to achieve.

1900
Charles Kingston
(1850–1908)

Constitution Bill

Charles Kingston was a lawyer and from 1893 to 1899 he was the Premier of South Australia. He helped to write the first draft of the Australian Constitution as a delegate to the Constitutional Conventions of 1891, 1897 and 1898. In 1900, Kingston travelled to London with Edmund Barton and Alfred Deakin to present the Constitution to British parliament so they could pass a bill to make the Constitution legal. Australia could not be federated if this did not happen.

Kingston was appointed by Edmund Barton as the Minister for Trade and Customs in the first Federal Government.

1901
Sir Edmund Barton
(1849–1920)

First Prime Minister

Sir Edmund Barton was a politician and barrister from New South Wales. After Henry Parkes died in 1896, Barton took over the push for Federation and gave speeches all over New South Wales. He used the catch cry: "A nation for a continent and a continent for a nation." Barton urged other politicians that the 'territorial rights' of the colonies should not be removed and that there should be free trade once Federation was established. As a delegate to the convention in Sydney, Barton helped draft the Constitution. He also declared his desire for the Federal capital to be in New South Wales; however, his modifications were defeated in parliament.

Barton became the first Prime Minister of Australia on 1 January 1901. Barton resigned in 1903 to become a judge of the High Court of Australia.

1903
Alfred Deakin
(1856–1919)

Second Prime Minister

Alfred Deakin was a Victorian politician and barrister who was a major supporter of Federation. He attended federal conferences and conventions and he travelled around the country giving speeches. Deakin wanted to have a weak Senate in the Federal Parliament. He wanted to limit their control over money and taxation and instead make the House of Representatives more powerful.

Deakin was appointed Attorney-General of the first Federal Parliament. He became the second Prime Minister of Australia after Edmund Barton resigned in 1903. He served as Prime Minister two more times and was a strong supporter of the White Australia policy.

Source: Australian History Centres: *Upper Primary*, p. 47, Blake Education; *Blake's Australian History Guide*, pp. 68–77, Pascal Press; Go Facts Australia: *Federation*, pp. 10–11, 22, Blake Education

TARGETING HASS 6 © PASCAL PRESS ISBN: 9781925726077

Research

Find out why Canberra became Australia's capital city.

Questioning

Write three questions you would like answered after reading about these important Australians. Can you find the answers?

1

2

3

Analysing

What were the differences in the viewpoints of these four important Australians in terms of their plans for a Federated Australia?

Communicating

Imagine you are Sir Edmund Barton, trying to convince parliament that the new Australian capital should be Sydney.

Write a short speech outlining your ideas and reasons. Present it to your class.

Evaluating and Reflecting

Was making Canberra the capital city of Australia a good idea?
Would a different location have been better? Give reasons for your answer.

The Australian Constitution

Before Australia could become a federated nation, a constitution had to be written. The Australian Constitution and the constitutions of many countries around the world are based on a document written 800 years ago called the Magna Carta (Latin for 'The Great Charter'). It is the first document to state that all men had to abide by the law, even the King. This is the main principle of the constitutions of Great Britain, the United States of America and Australia.

HISTORY

Magna Carta

The Magna Carta, signed by King John of England, was the first time there was a list of rules regarding how a country would be governed. It included 63 laws, but only three of the original laws remain in practice today. The most famous one states that every man has the right to a fair trial by jury. This idea forms the basis of our legal system today.

Legal Document

Australia's Constitution is the most important legal document in Australia. It contains rules that set out how Australia should be run. It explains how the parliament is to work, how laws are made and how the courts work. It also says what the Australian government and the state and federal parliaments can and can't do. The Constitution gives the federal government responsibility for running a federal parliament, trade, foreign affairs, taxation, postal and communication services, defence, immigration, currency, banking, social welfare and quarantine. Other aspects of Australian life, such as education, health, police and transportation, are the responsibility of state governments.

Constitutional Conventions

The people who wrote the Constitution knew how important it was to get it right. Representatives from each colony came together at meetings called constitutional conventions where the Constitution was discussed, drafted and rewritten. At the meetings, they looked at the constitutions of other countries, particularly those of Great Britain and America. They adapted the ideas they thought would work best for Australia and put them together in one document. Delegates at these conventions decided that a federation would need a new court, the High Court of Australia. It would hear challenges from governments, organisations or individuals who disagreed with the way a federal government acted under the Australian Constitution.

Referendums

The people of Australia voted on the draft Constitution in a series of referendums held between 1898 and 1900. For the Constitution to be accepted, a majority of voters in each colony had to agree. But unlike today, voting was not compulsory in 1898–99 so many people did not vote. Only South Australian and Western Australian women could vote. Indigenous Australians, Asians, Africans and Pacific Islanders were not allowed to vote in Queensland or Western Australia unless they owned property.

By July 1900, the majority of people had voted 'yes'. Yet before the Constitution could be made legal, it had to be approved by the British parliament. In 1900, British Parliament passed the *Commonwealth of Australia Constitution Act* which meant that Australia could finally become a nation. More than 100 years later, the Constitution is still much the same.

3

PART III.—The House of Representatives.
PART IV.—Provisions relating to both Houses.
PART V.—Powers of the Parliament:
CHAPTER II.—The Executive Government:
CHAPTER III.—The Federal Judicature:
CHAPTER IV.—Finance and Trade:
CHAPTER V.—The States:
CHAPTER VI.—New States:
CHAPTER VII.—Miscellaneous:
CHAPTER VIII.—Amendment of the Constitution.

CHAPTER I.
THE PARLIAMENT.
PART I.—GENERAL.

1. The legislative powers of the Commonwealth shall be vested in a Federal Parliament, which shall consist of Her Majesty, a Senate, and a House of Representatives, and which is hereinafter called "The Parliament." Legislative powers.

2. The Queen may, from time to time, appoint a Governor-General, who shall be Her Majesty's representative in the Commonwealth, and who shall have and may exercise in the Commonwealth during the Queen's pleasure, and subject to the provisions of this Constitution, such powers and functions ~~as~~ the Queen may think fit to assign to him. Governor-General.

3. The annual salary of the Governor-General shall be ~~fixed by the Parliament from time to time, but shall not be less than~~ Ten Thousand Pounds, and shall be payable to the Queen out of the Consolidated Revenue Fund of the Commonwealth. ~~The salary of a Governor-General shall not be diminished during his continuance in office.~~ Salary of Governor-General.

4. The provisions of this Constitution relating to the Governor-General extend and apply to the Governor-General for the time being or other the Chief Executive Officer or Administrator of the Government of the Commonwealth, by whatever title he is designated. Application of provisions relating to Governor-General.

5. Every member of the Senate, and every member of the House of Representatives, shall before taking his seat therein make and subscribe before the Governor-General, or some person authorised by him, an oath or affirmation of allegiance in the form set forth in the Schedule to this Constitution. Oath of allegiance. Schedule.

6. The

1891 draft of the Australian Constitution with corrections and notes made by Edmund Barton

	Great Britain	Australia	America
Has a constitution	✓	✓	✓
Two houses of parliament (called congress in America)	✓ House of Commons, House of Lords	✓ House of Representatives, The Senate	✓ House of Representatives, The Senate
Three levels of government	✗ Only 2 levels: National, Local	✓ National, State, Local	✓ National, State, Local
Is a republic (no king or queen)	✗	✗	✓
Is a monarchy (has a king or queen)	✓	✓	✗
Has a prime minister who is head of government	✓	✓	✗
Has a monarch who is head of state	✓	✓	✗
Has a president who is head of government and head of state	✗	✗	✓

Source: Australian History Centres: *Upper Primary*, p. 43, Blake Education; *Blake's Australian History Guide*, p. 70, Pascal Press; Go Facts Australia: *Federation*, p. 16, Blake Education

TARGETING HASS 6 © PASCAL PRESS ISBN: 9781925726077

Research

Research the constitution of *either* Great Britain *or* the United States of America.
Write five interesting facts about it.

1 ______

2 ______

3 ______

4 ______

5 ______

Questioning

What are the meanings of these words, as they are used in the text?

- jury ______
- constitution ______
- delegate ______
- referendum ______
- Act ______

Analysing

What does the table in the text tell you about the relationship between the Australian Constitution and the constitutions of Great Britain and America?

______

Communicating

The Constitution has rules that set out how Australia should be run. If you were to create a Constitution for your class, what would it say?

Evaluating and Reflecting

The Australian Constitution is over 100 years old and it is still much the same today as it was in 1901. Do you think the Constitution should be updated or changed? Why?

______

Indigenous Rights

The Constitution

Aboriginal peoples were only mentioned twice in the 1901 Australian Constitution and Torres Strait Islander peoples were not mentioned at all. Indigenous peoples were also not counted in the census and were often viewed as second-class citizens, so did not receive the same rights and freedoms as other Australians. They were even looked after by the same department that dealt with national parks, flora and fauna.

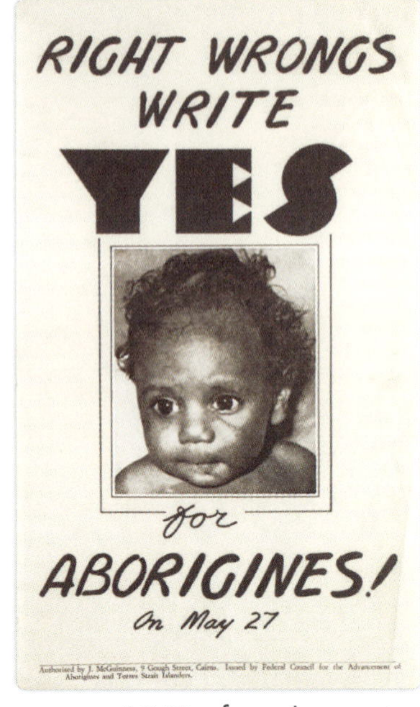

1967 referendum poster

Full Control

Only the newly formed states could make laws that applied to Aboriginal peoples. As colonies, most states had Aboriginal Protection Boards and roles such as Protector of Aborigines. These were meant to protect Aboriginal peoples, but actually controlled their lives. They were required to ask the Protector for permission to marry, move, travel and even to access their wages. The states controlled whom they could work for, how much they got paid, their education, housing and more.

Citizenship

In 1948, the *Nationality and Citizenship Act* was passed. This Act meant that all people born in Australia were classified as Australian citizens, including Aboriginal and Torres Strait Islander peoples. Anyone who was an Australian citizen was also considered a British subject.

Right to Vote

Although they were considered Australian citizens, Indigenous peoples still did not have the right to vote. In 1961, a Commonwealth Parliamentary Committee investigated and reported on Indigenous voting rights. This resulted in amendments to the *Commonwealth Electoral Act* in 1962, which allowed Indigenous peoples to enrol to vote in federal elections if they wished. It was not compulsory for Indigenous peoples to enrol to vote, unlike other Australians.

'Aboriginal Embassy' demonstration, 1972

Referendum

In 1967, a national referendum was held. Australians were asked if they agreed that Indigenous peoples should be counted in the census and that the Commonwealth Government could pass laws for Indigenous peoples. More than 90 per cent of Australians voted 'yes'. The Constitution was amended to include these changes.

Source: *Blake's Australian History Guide*, p. 77, Pascal Press

TARGETING HASS 6 © PASCAL PRESS ISBN: 9781925726077

Research

In 1972, a group of Aboriginal people set up a 'Tent Embassy' outside Parliament House in Canberra. Research the Tent Embassy and what it represented.

Questioning

1 Why didn't Aboriginal people receive the same rights and freedoms as other Australians?

2 How did the states control the lives of Aboriginal people?

3 What was important about the 1948 *Nationality and Citizenship Act*?

4 Why was the Constitution amended in 1967?

Analysing

What is the impact of the image of the young Aboriginal child in the 1967 referendum poster shown in the text?

Communicating

Use the information in the text to create a timeline of major milestones in the history of Indigenous voting rights. You can add additional information from your research.

Evaluating and Reflecting

In 1962, when Aboriginal people were allowed to enrol and vote in federal elections, do you think it should have been compulsory as it was for other Australians? Give reasons for your response.

Labourers and the White Australia Policy

HISTORY

EARLY MIGRATION

In the late 1700s and early 1800s, most of the migrants to Australia were either transported convicts or voluntary settlers. It was not until the gold rush of the 1850s that large numbers of people from all over the world came to Australia to try and strike it rich. Between 1851 and 1861, over 600 000 people came to Australia from overseas. After the gold rush ended, many chose to settle here permanently rather than return home.

CHINESE GOLD DIGGERS

About 42 000 of those who came to Australia to seek their fortunes in gold were from China. Many of them were under contract to agents and had to repay the costs of their voyages. They also sent money home to their families in China who lived in poverty. The Chinese gold diggers experienced racism and prejudice from white diggers. Riots and violence broke out at the camps, which resulted in the governments of Victoria and New South Wales placing restrictions on Chinese immigration.

Newspaper cartoon from 1856 that portrays the negative attitude towards Chinese gold diggers

SOUTH SEA ISLANDERS

Beginning in 1863, as many as 62 000 South Sea Islanders were brought to Australia over the next 40 years to work on Queensland sugar plantations. They mostly came from Vanuatu and the Solomon Islands, and were referred to as 'Kanakas' at the time — a word that is now considered offensive. Many men were kidnapped from their homes and forced into indentured labour in Australia. This was known as 'blackbirding'.

South Sea Islanders working on Queensland sugar plantations

Poster for a patriotic play, 1909

WHITE AUSTRALIA POLICY

White people in Australia became increasingly resentful of non-whites like the Chinese and South Sea Islanders, who usually worked for very low wages. They began to fear that migrants were threatening their jobs. After Federation in 1901, Parliament passed the *Pacific Island Labourers Act*. Between the years of 1904 and 1908, most of the remaining South Sea Islanders in Australia were deported. The government passed the *Immigration Restriction Act* in 1901, which is commonly known as the White Australia Policy. One of the main requirements of the Act was a dictation test that the person applying for immigration had to pass. The test was often in a language the applicant couldn't understand. This prevented many Asians and other non-Europeans from migrating to Australia.

Source: *Blake's Australian History Guide*, p. 92, Pascal Press

TARGETING HASS 6 © PASCAL PRESS ISBN: 9781925726077

Research

Investigate the *Immigration Restriction Act* 1901. Write five interesting facts about it.

1 ______________________________

____

____

2 ______________________________

____

____

3 ______________________________

____

____

4 ______________________________

____

____

5 ______________________________

____

____

Questioning

True or false?

1 Most immigrants chose to leave Australia after the gold rush.

____

2 Chinese migration was restricted because many of them were under contract to agents.

____

3 South Sea Islanders worked on sugar plantations.

____

4 White Australians didn't want South Sea Islanders taking their jobs.

____

5 The *Immigration Restriction Act* discriminated against Asians and non-European people.

____

6 Many immigrants came to Australia because the wages were low.

____

Analysing

How does the 1856 cartoon about Chinese gold diggers illustrate the negative attitude towards this group?

____

____

____

____

____

Communicating

Create your own 50 word dictation test for newly arriving immigrants to Australia.

Evaluating and Reflecting

How would you feel if you had to pass a test in a language that you hadn't studied and didn't understand? Explain your response.

____

____

____

____

UNIT 7

Strikes and Suffrage

HISTORY

The Right to Vote

The word 'democracy' comes from an ancient Greek word meaning 'rule of the people'. Australia is a representative democracy, which means that the Australian public votes for who they want to represent them in government. 'Suffrage' means the right to vote in elections. This is one of the most important rights in a democracy. When Australia first became a nation in 1901, women and Indigenous peoples were not allowed to vote.

Women's Suffrage Society

In the late nineteenth century, women throughout Europe and America were fighting for their right to vote. These women were referred to as 'suffragettes'. The first women's rights organisation in Australia was formed in 1884. Based in Melbourne, the Victorian Women's Suffrage Society was made up of both men and women. In the 1880s and 1890s, suffrage groups continued to be established until they were represented in each Australian colony.

The Womanhood Suffrage League of NSW, 1902

Women's Rights in Australia

The determination of the women in the suffrage movement led to changes across the world. In 1893, New Zealand was the first nation to give women the right to vote. The next year, South Australia gave women both the right to vote and to stand for parliament. Western Australia followed in 1899 by giving women the right to vote. The Australian Government passed the *Commonwealth Franchise Act* in 1902. This Act gave non-Indigenous women the right to vote in federal elections and to stand for Federal Parliament. Over the next six years, other states changed their legislation to recognise women's rights. By 1908, non-Indigenous women in all Australian states had been given the right to vote in state elections.

Women in Queensland vote in a state election for the first time, 1907

Equality at Work

Many men at the time believed that a woman's place was in the home, tending to domestic duties. The limited number of women who had jobs outside the home received much lower wages than men. Many working women fought for their rights by going on strike. The first of these strikes in Australia was staged in 1827 at a female factory in Parramatta. Convict women went on strike when their tea and sugar rations were taken away. They agreed to return when their rations were returned. In 1907, the Commonwealth Court of Conciliation and Arbitration declared that women's wages were to be 54 per cent of the basic male wage. This was the first step towards wage equality for women.

In 1891, the signatures of close to 30 000 women were collected in Victoria on a petition seeking their right to vote. Learn more about the 'Monster Petition' at the Parliament of Victoria website: www.parliament.vic.gov.au/about/the-history-of-parliament/womens-suffrage-petition.

Source: *Blake's Australian History Guide*, p. 76, Pascal Press

TARGETING HASS 6 © PASCAL PRESS ISBN: 9781925726077

Research

Investigate the contributions made by Henrietta Dugdale and the Victorian Women's Suffrage Society to the women's suffrage movement in Australia.

____

____

____

____

____

____

____

____

Questioning

Write four questions you would like answered after reading the text.

1 ______________________________________

____

____

2 ______________________________________

____

____

3 ______________________________________

____

____

4 ______________________________________

____

____

Analysing

How has the role of women changed since the 1800s?

Role of women in the 1800s	Role of women in modern society

Communicating

Choose something in society that you think should be changed. Is it something at home, at school, in your community?

Create a poster to petition for that change to happen.

Evaluating and Reflecting

What do you think were some of the advantages and disadvantages caused by the women's suffrage movement in Australia? Explain your response.

____

____

____

____

____

Spanish Influenza

INFLUENZA

The word **influenza** comes from 15th century Italy. People of the time believed that the stars 'influenced' diseases and infections. Influenza is a virus which causes upper respiratory tract infections. It mutates rapidly and constantly, so people cannot develop a long-lasting immunity to it.

THE SPANISH FLU

In 1918 at the end of World War I, a virulent strain of influenza known as the 'Spanish flu' swept through Europe. The virus did not originate in Spain but the first widespread reports of infection and transmission occurred there.

The Spanish flu arrived in Australia, accompanying servicemen returning from the war. The first reported case was a former soldier from New South Wales. Quarantine measures relied on the medical officers of ships docking in Australian ports to admit to having cases of Spanish influenza on board, or for returned soldiers to self-report their illness and be isolated from the community. The virus quickly escaped these measures.

The Spanish influenza pandemic was widespread. Approximately one-third of the world's population became infected, resulting in around 50 million deaths. In 1919, Australia's population was 5 million. Forty percent of the population contracted Spanish influenza and approximately 15 000 died as a result.

Indigenous communities were affected particularly badly. In Queensland at the Barambah government reserve for Aboriginal people, 590 contracted Spanish influenza. Within three weeks, 90 had died. Australian medical practitioners called it 'pneumonic influenza' because of the way the virus attacked the lungs and respiratory systems of its victims.

Australia had time to plan and make precautions for the pandemic due to its distance from other countries, particularly those in Europe. The federal government was given the power to close borders between states, and state health officers had to notify each other of cases of the Spanish flu. But there was much division and states held off informing authorities of suspected flu cases. Also, there were not enough doctors and nurses to treat infected people because many were still overseas after serving during World War I.

As the virus spread, individual states made their own decisions. Some closed schools. Others turned schools into makeshift hospitals for the sick. Western Australia stopped the trans-continental train and isolated all of its passengers. Tasmania imposed strict quarantine measures and as a result had the lowest rate of infection and death in the country. But the island state also suffered greater economic hardship. They ran short of goods such as flour and their tourism industry was badly affected.

TREATMENT AND CURE

The most effective way of dealing with the illness was prevention. There were public health campaigns involving social distancing, the wearing of face masks and the importance of washing hands. Gradually the infection rates slowed due to a combination of good hygiene, quarantine procedures and a vaccine produced by Australia's Commonwealth Serum Laboratories.

COVID-19

There are similarities between the 1919 Spanish Influenza pandemic and the 2020 COVID-19 pandemic. Both created a health crisis as well as an economic crisis around the world. Across Australia, states once again closed their borders and Tasmania quarantined itself. Anyone travelling interstate had to self-isolate on arrival for 14 days. Cinemas, playgrounds and churches were closed, sporting events and competitions were cancelled and both pandemics resulted in the cancellation of Sydney's Royal Easter Show.

Illustrator: Paul Lennon

HISTORY

Research

Investigate pandemics. What are they? How often do they occur?

Questioning

1 Why is the virus called 'influenza'?

2 Why did the 1919 Spanish flu spread so easily around the world?

3 How can the spread of the influenza virus be slowed?

Analysing

Compare the 1919 Spanish influenza pandemic with the 2020 COVID-19 pandemic.

Similarities	Differences

Communicating

Social distancing and good personal hygiene are the keys to minimising the impact of a viral pandemic.

Create a TV commercial, radio advertisement or poster to encourage people to 'stay safe' during a pandemic.

Evaluating and Reflecting

During a pandemic, do you think individual people should be able to choose to self-isolate if they feel unwell, or should the government have control over individual freedoms for the good of the whole community? Give reasons to support your opinion.

Reasons for Migration in the 20th Century

The word *migrate* means 'to move from one place to another'. During the 20th century, people migrated to and from Australia for many different reasons. These reasons are called push and pull factors.

Events that 'pushed' people to Australia in the 1900s

Wars: Many people were pushed out of their homelands in Europe because of World War II. After the war, they were unable or unwilling to return to their homelands and were forced to live in refugee camps. These people became known as 'displaced persons'. Australia wanted to increase its population, so the government agreed to resettle many of these people.

Religious Persecution: People who are forced out of their countries because of their religion are called religious refugees. Jewish people were pushed out of Germany in the 1930s and 1940s because of unfair and poor treatment, known as religious persecution. Many were forced to flee, often with no money and no possessions. Australia agreed to accept some of these religious refugees.

Laws and Government Policies: Sometimes governments make laws that make life difficult for people. After Federation in 1901, the government created laws known as the White Australia Policy. These were meant to create a mostly white and British population. Many non-British people were pushed out of Australia through deportation.

Push factors are things that 'push' people out of a country, while pull factors are things that 'pull' people towards a country. These factors are often significant events, such as wars, or laws and policies made by a country's government.

Events that 'pulled' people to Australia in the 1900s

Work opportunities: The Snowy Mountains Scheme in New South Wales was a project that needed a very large workforce. Australia did not have many people with the required skills to build the project. Many experienced migrants from Europe were pulled to Australia to work on the scheme, as well as those who came because they were drawn by the high wages.

Better Quality of Life: Many people believed Australia offered them a better way of life than that of their own countries. As Australia was a young and developing country, many people, particularly those from Britain and Ireland, were pulled to Australia by the warm weather and the possibilities of better jobs and owning a home.

Laws and Government Policies: Sometimes governments make laws and policies that encourage certain people to migrate to a country. Although the White Australia Policy pushed many non-British people out of Australia, it also pulled people from the United Kingdom and Ireland to the country. The government also provided cheap passage on ships and services to help people settle when they arrived.

Source: Australian History Centres: *Upper Primary*, p. 61, Blake Education

TARGETING HASS 6 © PASCAL PRESS ISBN: 9781925726077

Research

What was the Snowy Mountain Scheme? How was it important to the development of Australia?

Questioning

Write three questions to evaluate a classmate's understanding of the text. How well did they do?

1

2

3

Analysing

Complete the table below to examine how different groups would have viewed the White Australia Policy.

Person already living in Australia	Non-European migrant	British migrant

Communicating

Where would you like to live? List some factors that 'push' you there and some factors that 'pull' you there.

Evaluating and Reflecting

The White Australia Policy 'pulled' many people from the United Kingdom and Ireland at a time when Australia needed more people. Does this mean that the policy was a good one? Give reasons to explain your answer.

Vietnam War and Refugees

Before 1975, there were only a few hundred people with a Vietnamese background living in Australia. The end of the Vietnam War changed that.

From the early 1960s until 1975, the Vietnam War raged between the communist forces of North Vietnam (supported mainly by China) and the anti-communist forces of South Vietnam (supported by the United States and several other countries, including Australia). The war was long and difficult. When the North captured the capital of South Vietnam, it was considered the end of the war. Many people fled South Vietnam because they did not want to live under communist rule and believed they would be punished or killed for fighting against the north.

More than two million people made their way to refugee camps in Thailand, Hong Kong, Malaysia and Indonesia. As the camps in these countries became more crowded, the Australian Government agreed to accept some refugees from South Vietnam. Two plane loads of orphans - mostly babies - arrived in April 1975. Australian officials visited a refugee camp in Hong Kong and chose another 201 refugees to settle in Australia.

Tan Thanh Lu arrived in Australia on a fishing boat similar to this one.

The following year, boatloads of South Vietnamese travelling in overcrowded and leaking fishing boats made their way to Australia, becoming Australia's first boat people. Most came through Malaysia and other Asian countries which had turned them away. As the number of boats increased (54 within three years), the initial sympathy Australians had for the refugees changed. Unemployment in Australia was growing and people feared that refugees would take their jobs. One trade union went on strike to protest against the special treatment they believed Vietnamese refugees were getting.

Ten years after the war ended, more than 90 000 refugees from Vietnam, Cambodia and Laos had arrived in Australia. These were the first large groups of Asian people to migrate to Australia since the Chinese gold rushes in the mid 1800s.

Vietnam has been the sixth-largest source of immigration to Australia, behind the United Kingdom, New Zealand, China, India and Italy. By 2018, according to the Australian Bureau of Statistics, there were 256 000 Australians who had been born in Vietnam.

Among the refugees escaping war-torn Vietnam was Anh Do, the now famous Australian author, actor and artist. He and his family survived, despite terrifying storms and attacks by pirates. His book, *The Little Refugee,* is the story of his journey to his new home and his life growing up in Australia as a boy with no English and 'funny' lunches. But along the way, Ahn learned to adapt and look at the bright side of life. Many of his personal experiences are reflected in his work as an author and comedian.

Source: Australian History Centres: *Upper Primary*, p. 67, Blake Education; Go Facts Australia: *Migration*, pp. 24–5, Blake Education

TARGETING HASS 6 © PASCAL PRESS ISBN: 9781925726077

Research

Investigate Tan Thanh Lu and the fishing boat, the *Tu Do*.

Questioning

Imagine you could interview a refugee to Australia. What would you ask them about their experiences? Write four questions.

1

2

3

4

Analysing

Examine the photograph of the Vietnamese fishing boat. What challenges do you think refugees faced when travelling on boats such as these to Australia?

Communicating

Create a map to show the route the *Tu Do* took on its journey from Vietnam to Australia.

Evaluating and Reflecting

Should the experiences of refugees coming to Australia be shared with and taught to students in schools today? Are these personal stories important? Explain your point of view.

UNIT 11

Notable Australians

HISTORY

> **Did you know?**
> Twelve Australians have won Nobel Prizes, in the fields of literature, physics, chemistry, and physiology or medicine.

Anyone can contribute to the richness of life in their country – you don't need to be 'famous' to do this. Most people don't seek greatness, nor do they achieve it overnight. The amazing discoveries and achievements of these Australians have one quality in common: they result from hard work. Each person, in their own way, has enriched and shaped the nation and the world. They are role models who challenge us to persevere and excel in whatever we choose to do.

Quentin Bryce
1942–

Quentin Bryce was Australia's first woman Governor-General. She is also a lawyer, educator and human rights advocate.

Quentin was born in Brisbane and, in her own words, spent her childhood "in a little bush town". In 1965 she graduated from the University of Queensland with degrees in arts and law. She was one of the first female barristers in Queensland.

Throughout her career, Quentin strived to improve the lives of women, families and young people. She had leadership roles with the National Women's Advisory Council, Queensland Human Rights and Equal Opportunity Commission, and the Sex Discrimination Commission.

In 2008 Quentin became the twenty-fifth Governor-General of Australia. Her role was to protect the Australian Constitution and to enable the work of the Commonwealth parliament and government. "I promise to be alive, open, responsive, and faithful to the contemporary thinking and working of Australian society." On taking office, Quentin stated one of her goals was to focus on protecting the rights of the country's Indigenous peoples.

Quentin retired as Governor-General in March 2014.

Charles Perkins
1936–2000

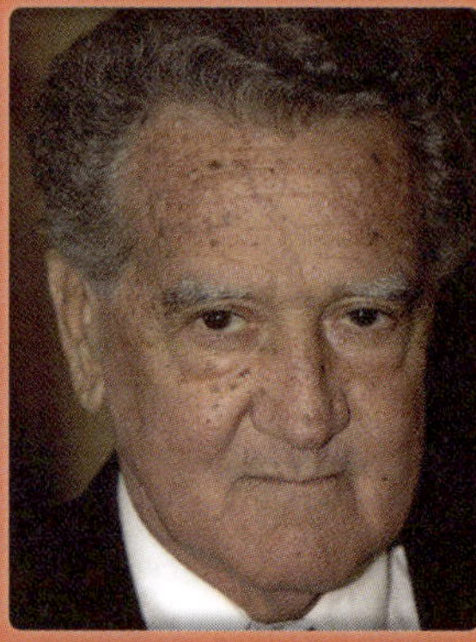

Charles Perkins was a courageous and outspoken activist for the rights of Aboriginal people.

Charles was born at the Alice Springs Telegraph Station Aboriginal Reserve. When he was 10, he was taken to Adelaide to be educated. After finishing school in 1957, his talent as a soccer player took him to England to play with Everton Football Club. When he returned to Australia, Charles attended the University of Sydney and in 1965 became the first Aboriginal to graduate from university.

All his life, Charles had seen that Aboriginal people were treated differently. He decided that this had to change. In 1965 he organised what became known as the Freedom Ride. He took 30 university students on a bus tour of towns in New South Wales. The group protested outside swimming pools, movie theatres and clubs that refused entry to Aboriginals as "a new way of promoting a rapid change in racial attitudes in Australia". This was the first of many actions that led to the successful 1967 national referendum on Indigenous rights.

From 1969, Charles worked to improve Aboriginal schooling, employment, housing and health. He was the head of the Department of Aboriginal Affairs 1981-1988.

Ian Frazer,
1953–

Ian Frazer works to develop vaccines that protect people from cancer.

Ian was born in Scotland. He was a curious child who liked to take things apart and then put them back together again. At university he planned to study physics but changed his mind and became a medical doctor.

In 1980, Ian immigrated to Australia. His area of research was immunology – how the body protects itself from disease. After working in Melbourne, he moved to Queensland to study and teach at the University of Queensland. Here Ian began working with a Chinese-born researcher named Jian Zhou, and a team of scientists. After many years of work, they developed a vaccine that destroys viruses which can cause cervical cancer in women. They did this by creating particles that mimic the virus. This was a major achievement because this cancer was a leading cause of cancer deaths for women.

The new vaccine was released in 2006. Since then, more than 40 million women and girls around the world have received it.

Ian received the Australian of the Year award in 2006. Speaking of this award, he said, "It's nice to get the recognition for science. I think it's very important to show the community at large that science contributes to society." Ian is now working on a vaccine for skin cancer.

Source: Go Facts Australia: *Great Australians*, pp. 8–11, 22–3, Blake Education

Research

May Gibbs, Albert Namatjira and John Flynn are also great Australians. Choose one person, and investigate their contribution to Australian society.

Questioning

Write four questions you would ask the person you researched about their life and their legacy.

1 _____

2 _____

3 _____

4 _____

Analysing

How are these notable Australians similar?

Communicating

Who else do you think should be considered a 'Notable Australian'? Why? Are they an artist, scientist, sportsperson or community volunteer?

Write a two-minute speech to convince your audience to agree with your choice.

Evaluating and Reflecting

Of the three Australians in the text – Quentin Bryce, Charles Perkins and Ian Frazer – who do you think has made the most significant contribution to Australian society? Explain your response.

UNIT 12

A Diverse Land

HISTORY

'Multicultural' describes a society in which people from many different cultures live together in harmony. Australia is a multicultural country and will become even more so in the future.

Since British colonisation, there have been many debates about whether Australia should encourage immigration and be multicultural.

Those that oppose it sometimes say that migration from non-European countries is changing the face of Australia. They believe that a stable, cohesive society is not possible if there are many cultures that don't share values, history and language. Some people believe that immigration – from any country – puts too much pressure on the natural environment. Australia simply can't support a large population.

Those who favour a multicultural Australia argue that migrants improve and grow Australia's economy. More houses and schools are built. More goods and services need to be supplied. Australia needs migration in order to grow its population and prosper. Since 2005, net migration has overtaken natural increase (births minus deaths) as the main source of population growth.

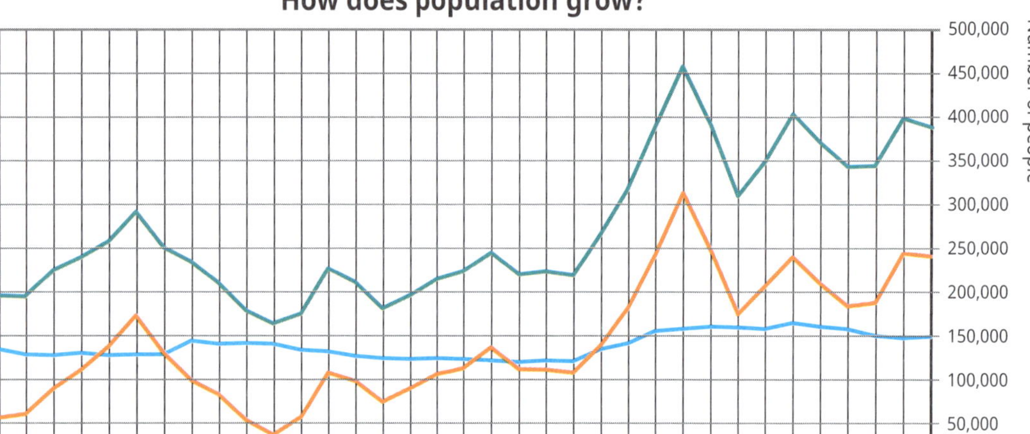

Source: Australian Bureau of Statistics

When migrants travel to Australia, they bring some of the customs and traditions of their home countries. They have introduced new food, clothes, languages, religions, music, art, sports and games. Today, many of these traditions and customs have become part of the everyday lives of all Australians, regardless of their heritage.

Bendigo Easter Festival
Many Chinese people came to Victoria during the gold rush in the 1850s and 1860s. They have taken part in the Bendigo Easter Festival since 1871 and in 1879 they joined the procession. Each year, Sun Loong, the world's longest Chinese Imperial Dragon, is paraded as part of the Gala Procession. The Bendigo Chinese Association aims to promote good will and understanding between Chinese, Australians and people from different nationalities.

Auburn Gallipoli Mosque
This mosque is located in Sydney, New South Wales. Its construction was completed in 1999. Many people who worship here are from the local Turkish community. Its name refers to the legacy of the Gallipoli Campaign of World War I, which was a significant event for both Australian and Turkish people. The mosque provides language education and community and cultural activities.

Soccer
Soccer was played in Australia as far back as the 1830s. However, it wasn't until the end of World War II that the game became popular. Post-war migrants from Europe who settled in Australia brought their love of the game with them. Playing soccer brought communities and people of different ethnicities together. Many migrant groups, such as the Greeks and Italians, set up large soccer clubs.

Source: Go Facts Australia: *Migration*, pp. 24–5, Blake Education; Australian History Centres: *Upper Primary*, p. 73

Research

Select **one** cultural group that has made its home in multicultural Australia. How has it influenced Australia's sport, culture, cuisine, clothing and general way of life?

Questioning

Use the text to help you write questions that have these answers.

Answer: the natural environment
Question:

Answer: soccer
Question:

Answer: mosque
Question:

Answer: the economy
Question:

Analysing

Examine the graph showing how Australia's population has changed. What trends can you see? What might be influencing those trends?

Communicating

Using the cultural group you researched, create a collage to illustrate the contribution of this group to Australian society and the way we live.

Evaluating and Reflecting

Has multiculturalism had a positive impact on Australians and our way of life? Explain your answer, using examples.

UNIT 13

Asia

Asia is one of the world's most diverse continents. It has many cultures, each with its own food, art, music, religion, dance, traditions and language. It also has some of the richest and poorest countries on the planet. Many of the goods we buy are produced in Asia, making it important economically as well as culturally.

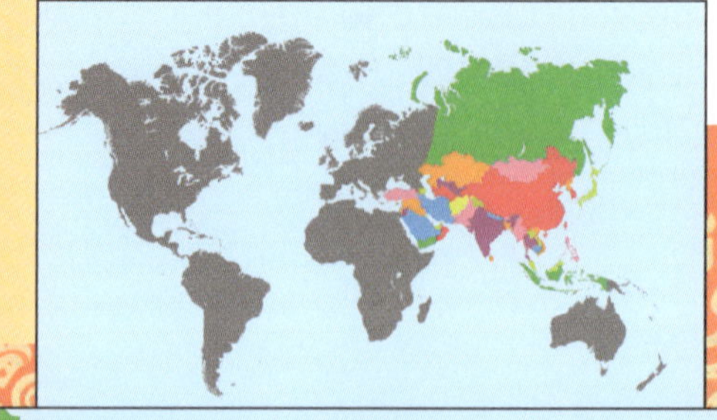

ASIA FACTS

- Sits in the Northern Hemisphere but also extends to the Southern Hemisphere
- Contains 48 countries
- The world's most populated continent (over 4 billion people); contains the two most populated countries in the world (China and India)
- The largest of the continents (more than five times larger than Australia); covers more than 30% of the Earth's land area
- The only continent that shares land borders with two other continents (Africa and Europe)

Environmental Characteristics

Asia is vast. It extends from the tropics in the south to the arctic region in the north. Because of its size, it contains a wide variety of environments and natural features.

Climate

Asia contains some of the wettest, driest, hottest and coldest places on Earth. Many climate zones are found in Asia, including semi-arid, arid, humid, subtropical and continental. In turn, many different environments from rainforests to deserts to snow-capped mountains are found there.

Natural Features

Landscapes

- Himalayan Mountains (Nepal, India, Bhutan)
- Tibetan Plateau (China)
- Mongolian Steppe (Mongolia)
- Gobi Desert (Mongolia)

Vegetation

- rainforests
- grasslands
- wetlands
- tundra forests
- coniferous forests
- savanna

Natural resources

- oil
- gas
- coal
- forests
- fisheries

Landforms

- Mount Everest (world's highest mountain, Nepal and China)
- Lake Baikal (world's deepest lake, Russia)
- Dead Sea (lowest land point in the world, Israel and Jordan)
- Yangtze River and Shilin Stone Forest (China)
- Ha Long Bay (Vietnam)
- Chocolate Hills (Philippines)
- Mount Bromo (volcano, Indonesia)

Source: Australian Geography Centres: *Upper Primary*, p. 31, Blake Education

Research

Asia has the two most populated countries in the world – China and India. Choose **one** of these countries and research the environmental impacts of their growing populations.

Questioning

Write five questions that have the five 'Asia Facts' as the answers.

1

2

3

4

5

Analysing

How does Asia's size and location impact its environmental characteristics?

Communicating

If you could travel to Asia, which landscape or landform would you most like to visit, and why?

Write a short paragraph presenting your point of view and the reasons for your choice.

Evaluating and Reflecting

What influence has Asia had on your own life? Think about family, food, religion, trade, travel and friends. Has that influence been positive or negative? Explain why.

TARGETING HASS 6 © PASCAL PRESS ISBN: 9781925726077

UNIT 14

Mongolia and Pakistan

COUNTRY	**MONGOLIA (East Asia)**  	**PAKISTAN (South Asia)**  
FACTS	**Capital:** Ulaanbaatar **Area:** 1 564 116 km^2 **Official language:** Mongolian **Major religion:** Buddhism **Population:** 3 million	**Capital:** Islamabad **Area:** 796 095 km^2 **Official languages:** English, Urdu **Major religion:** Islam **Population:** 193 million
CLIMATE	The climate in Mongolia is extreme. In one day, temperatures can range from 30 °C to 0 °C. It also has long, bitterly cold winters and short, hot summers. What little rainfall it receives falls in summer, and the country is generally dry.	Pakistan is in the warm temperate zone. It has hot, dry summers and cool to cold winters. In summer, some areas can reach up to 50 °C. It rains very little in Pakistan and much of the country is very dry.
NATURAL FEATURES	High mountain ranges dominate northern and western Mongolia, while the dry Gobi Desert dominates the south-east. The north-west region has over 300 lakes. Temperate grasslands called steppes cover much of Mongolia, and large rivers run through the country.	Pakistan is divided into three major regions: the northern mountains, the Indus River plains in the east and the Balochistan Plateau in the west. Glaciers are found in the mountains. The Indus River plains are fertile, but the rest of the country is dry, arid and rocky.
SETTLEMENTS	A third of the population lives in Ulaanbaatar. The rest live in rural areas. Many of these people are farmers herding goats, sheep, cattle and horses, or growing wheat and barley. Herders live a nomadic lifestyle, moving around to new grazing areas.	Most of the population lives in the Indus River plains. They are the most fertile areas in the country and where much of the farming occurs. More than 60% live in rural areas, but many also live in large cities.

Some Mongolians live in *gers*, which are easy to move from place to place.

Many Mongolians make a living by farming livestock such as sheep.

Some of the world's highest mountains are found in Pakistan, including the world's second highest mountain, K2.

Many people in Pakistan live in cities like Lahore in the fertile east of the country.

Source: Australian Geography Centres: *Upper Primary*, p. 35, Blake Education

TARGETING HASS 6 © PASCAL PRESS ISBN: 9781925726077

Research

Find out the location of the capital cities of Mongolia and Pakistan, in longitude and latitude.

What is the distance between these two capital cities, in kilometres?

Analysing

Use the Venn Diagram to compare and contrast Mongolia and Pakistan.

Mongolia

Pakistan

Questioning

Use the text to help you write five quiz questions. Test a friend with your questions. Do they know the answers?

1

2

3

4

5

Communicating

What would it be like to live in a *gers*? What benefits and drawbacks would they have?

Discuss this in a small group and use the conversation to create a PMI (positive, minus, interesting) chart.

Evaluating and Reflecting

What problems might Pakistan face in the future if most of its population live in the fertile east of the country? Consider issues such as food and water supplies and the impact of population density.

Environments

From tropical rainforests to deserts, Asia's environments are just as diverse as its people.

Rainforest Environments

Most of South-East Asia is situated within the tropics. It has a warm to hot and humid climate, with wet and dry seasons. More than half of Brunei is covered with some of the oldest rainforests in the world. They have two types of rainforest: hill forests which are drier and found in higher altitudes, and swamp forests which are affected by tides and local flooding. Brunei rainforests have high levels of biodiversity and contain the rare proboscis monkey, clouded leopard and sun bear. Their habitats benefit from the low rate of deforestation in Brunei.

Desert Environments

In North Asia, the Gobi is one of the world's largest deserts, covering vast areas of Mongolia and China. Most of the desert is stone, not sand. Temperatures can reach 45 degrees Celsius in summer but fall to -40 degrees Celsius in winter. The Gobi has fierce winds and sandstorms. They regularly blow sand across Asia, even as far as the United States of America. As a result of desertification, the Gobi Desert is expanding rapidly. With little soil cover, vegetation in the Gobi Desert is sparse.

Gobi sand dunes. 'Gobi' is Mongolian for 'waterless place'.

GEOGRAPHY

River Environments

The Mekong River runs through China, Myanmar, Laos, Thailand, Cambodia and Vietnam before reaching the South China Sea. About 24 million people – the equivalent population of Australia – live within five kilometres of the river. They rely on it for transportation, irrigation, food and water.

The Mekong River runs through six Asian countries.

The Mekong is one of the biggest inland fisheries in the world. Millions of fish breed in the river every year. There are three main reasons why this occurs:

- The river is long and free flowing. Many fish travel long distances to breed. They can do this because the Mekong has few obstacles, such as dams, blocking their path.
- A lot of fish food. Water flows into the Mekong through large areas of forests and floodplains. As it moves through these areas, it picks up nutrients and sediments that fish eat.
- Regular water flows. The fish in the river use the rainy seasons to move upstream to their breeding grounds.

The Mekong has huge potential as a source of renewable energy. Governments have proposed building more than a dozen dams on the lower section of the Mekong to generate hydroelectricity. However, such dams can seriously harm the environment and the 60 million people, many of them poor, who depend on the Mekong River. Dams will change the way the Mekong floods, which will affect how it provides food.

Buying and selling at a floating market

The Mekong is home to some of the world's poorest people and many actually live on the river.

Source: Go Facts Geography: *Asia*, pp. 14–5, Blake Education

TARGETING HASS 6 © PASCAL PRESS ISBN: 9781925726077

Research

The archipelago of Indonesia contains the world's third largest tropical forest. Find out about Indonesia's rainforests and the plants and animals that can be found there.

Questioning

Write four questions about living on the Mekong River that you would like answered after reading the text.

1

2

3

4

Analysing

Look at the image of the floating market.
Why do you think people buy and sell products in this way?

Communicating

Create a map to show the areas of rainforest on the islands of Indonesia. Share and discuss it with a partner.

Evaluating and Reflecting

The Mekong River is home to some of the world's poorest people and many actually live on the water. Why do you think this is so?

UNIT 16

Demographic Differences

Demographics are statistics about a country's population. Demographers (people who study demographics) look at different population characteristics, including size and density. Demographics help governments plan for the future. They are also useful for comparing countries around the world and for showing changes in populations over time.

Population density is worked out by dividing the number of people who live in a country by the size of the country. The answer shows how many people could live in each square kilometre of land if they were spread out evenly across the country. This indicates how crowded (densely populated) a country is.

Population size refers to how many people live in a country. This information is often shown as a population pyramid.

Developed Country Australia (2019)	
Size of country (km^2)	7 741 220
People per km^2	3.3

Developing Country India (2019)	
Size of country (km^2)	3 287 263
People per km^2	464

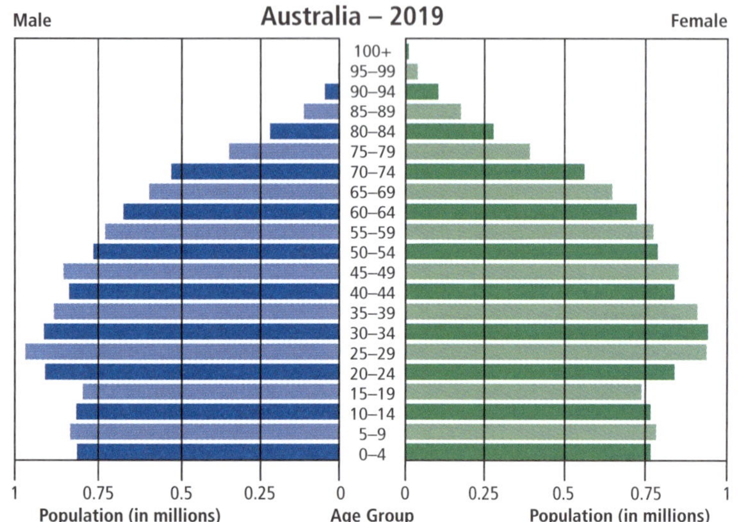

In 2019, the number of females aged 5–9:
- in Australia was approximately 780 000
- in India was approximately 54 750 000.

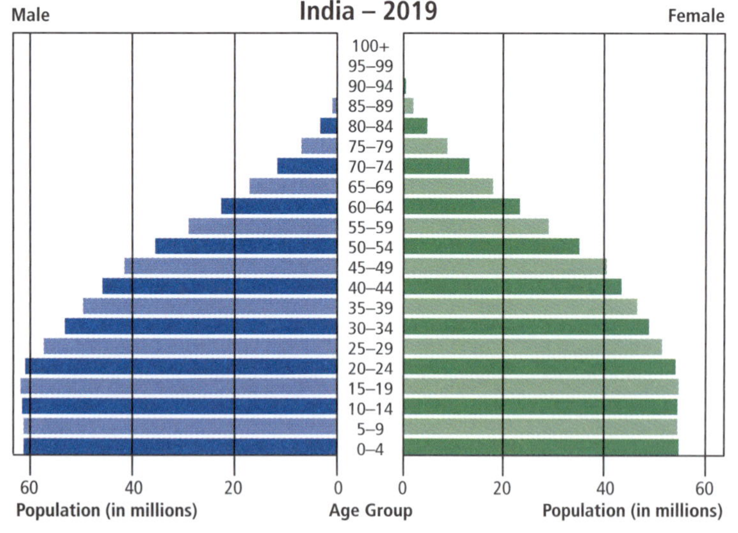

Demographics also include information about per capita income, health (life expectancy) and energy consumption (how much energy a population uses). When we study the demographics of different countries, we often see great differences. These differences make countries around the world diverse.

Health (life expectancy)
If people are healthy, they live longer. Studying the average life expectancy of a population gives an indication of how healthy the population is.

Energy consumption
Electricity generation, transport and industry use high levels of energy. As developed countries (countries with a high PCI) can access these more, they tend to use more energy than developing countries.

Per Capita Income (PCI)
The average income for each person in the population is called per capita income (PCI). The PCI is worked out by dividing the total amount of money made by the country by the number of people who live in that country.

Note: Australia's PCI in June 2019 was US $53 825. This doesn't mean that everyone in Australia earned that much money. It is an average of the whole population.

Life expectancy (2019)	Australia	USA	Norway	United Kingdom	Japan	Chile	Indonesia	Fiji
Male	81	76	81	80	81	77	69	66
Female	85	81	84	84	87	82	73	69

Source: Australian Geography Centres: *Upper Primary*, p. 41, p. 43, Blake Education

GEOGRAPHY

Research

Australia is a developed nation. India is a developing nation. What social and economic differences are there between developed and developing nations?

Developed nations	Developing nations

Questioning

Write four questions that can be answered by the life expectancy table. Remember to ask different types of questions: who, what, where.

1 ______________________________

____

____

2 ______________________________

____

____

3 ______________________________

____

____

4 ______________________________

____

____

Analysing

Most countries with lower Per Capita Incomes also have lower life expectancies. Why do you think this is?

____

____

____

____

____

____

Communicating

Imagine it is census time and you have been employed by the Australian government to collect demographic data from people living in Australia.

Write a short persuasive text to convince people to participate and give honest answers to your questions.

Evaluating and Reflecting

Governments study demographics to plan for the future. If a population pyramid shows a significant increase in births in a particular city or state, how would the government need to react? Why?

____

____

____

____

____

UNIT 17

Indigenous Peoples and the UNDRIP

Indigenous Peoples

A country's indigenous peoples are the descendants of the people who lived there before it was occupied or conquered by others. Since their lands have been occupied, many indigenous peoples have had difficult lives. They have been forced off their lands and denied many basic human rights, including the right to make decisions about things that affect them and the right to practise their cultures. When they have tried to assert their rights, many have been killed, imprisoned or tortured. Today, indigenous peoples make up only around 5% of the global population, but approximately one-third of the world's poorest people.

There are more than 370 million indigenous peoples in more than 90 countries, each with their own unique culture and languages. Australia's indigenous peoples are Aboriginal Australians and Torres Strait Islander peoples.

United Nations Declaration on the Rights of Indigenous Peoples (UNDRIP)

The United Nations is an international organisation. Over 190 countries around the world are members, and they meet regularly to discuss and make decisions about important issues that affect people around the world. On 13 September 2007, the United Nations General Assembly adopted (accepted) the United Nations Declaration on the Rights of Indigenous Peoples (UNDRIP). This is a document that explains how the rights of indigenous peoples should be protected. It contains 46 articles (statements) that describe these rights and provides ideas about how governments can protect these rights. It is hoped the declaration will help indigenous peoples and their cultures survive and thrive with dignity in an ever-changing world.

Indigenous Brazilian women and men

Yugambeh Aboriginal man

Akha women in Laos

Indigenous Fijian men

The UNDRIP covers four main themes. They are:

1. **The right of indigenous peoples to self-determination**
   Indigenous peoples should have the right to control their own lives by deciding what they think is best for them.
2. **The right to cultural identity**
   Indigenous peoples have the right to live and celebrate their own unique cultures.
3. **The right to be free from discrimination**
   All people should be treated equally regardless of culture, race, age or gender.
4. **The right to free, prior and informed consent**
   Before any decisions that affect them are made, indigenous peoples must be told what is proposed and they must freely (without any pressure from non-indigenous people) make up their own mind regarding whether they agree to the proposals or not.

Source: Australian Geography Centres: *Upper Primary*, p. 47, Blake Education

GEOGRAPHY

TARGETING HASS 6 © PASCAL PRESS ISBN: 9781925726077

Research

Who are the Akha people of Laos? Investigate them and write five interesting facts.

1 ______________________________

2 ______________________________

3 ______________________________

4 ______________________________

5 ______________________________

Questioning

What are the meanings of these terms?

- self-determination: ______________________________
- cultural identity: ______________________________
- informed consent: ______________________________
- declaration: ______________________________
- proposal: ______________________________

Analysing

When examining the issue of world exploration and global discovery, in what ways might the viewpoints of indigenous peoples from around the world be similar?

____

Communicating

Create a map of the world showing ten different indigenous peoples and their homelands.

You may need to do some extra research.

Evaluating and Reflecting

When the UNDRIP was first proposed in 2007, Australia was one of only four countries that voted against it. This position changed in 2009 and the government supported the Declaration. Was this change the right move? Should Australia support this Declaration? Explain your answer.

____

UNIT 18

Fukushima

GEOGRAPHY

On 11 March 2011, an earthquake struck off the coast of north-eastern Japan. It caused a tsunami that hit the coast and triggered a nuclear emergency.

The tsunami devastated a 670-kilometre stretch of coastline and powered on inland for kilometres. Almost 16 000 people died. Most towns had seawalls, but they proved useless against the scale of the surging water. The town of Rikuzentakata had tsunami shelters designed for waves 3–4 metres in height, but they were destroyed by a surge of water 13 metres high.

The earthquake rocked a nuclear power plant in Fukushima Prefecture and caused its three active reactors to shut down automatically. However, the tsunami also breached the plant's seawall and flooded its control systems. The reactors could no longer be cooled – they overheated and began to release radiation into the atmosphere.

Local authorities ordered everyone within two kilometres of the plant to evacuate. Over the coming days, the Japanese government widened the evacuation zone to 30 kilometres. In total, about 160 000 people were evacuated. Okuma is the town next to the nuclear power plant. By 2019, fewer than 400 of the original 10 000 residents had returned to live there.

Many Japanese people relied on the nuclear power plant for work. Most of them lost their jobs and their homes in the disaster. About 13 000 square kilometres of land was contaminated with radiation, which will take years to remove. Food produced in the prefecture and fish caught at sea were also contaminated. Many people – in Japan and other countries – remain suspicious about whether it is safe to eat.

Within days of the disaster, 91 countries offered aid to Japan, mainly search and rescue teams and medical teams. Australia sent 72 rescue workers and a cargo plane. The small nation of Maldives sent 86 400 cans of tuna as emergency food supplies.

Before the disaster, Japan generated about 30 per cent of its electricity from nuclear power plants. Following the disaster, it shut down all 48 of its reactors for compulsory safety checks. This meant that Japan had to import huge amounts of natural gas, coal and oil to generate electricity, at enormous cost. Australia is the largest supplier of gas to Japan and also supplies most of its coal.

Many towns along the coast, like Wakuya, were flattened by the tsunami.

The tsunami strikes the city of Miyako.

Public support for nuclear power fell in Japan and around the world.

Japan's nuclear emergency caused many people around the world to rethink their reliance on nuclear power. When asked about her response to the Fukushima incident, Australia's Prime Minister at the time, Julia Gillard, said that Australia would not be using nuclear energy. Instead, the focus would be on renewable energy sources such as solar and wind.

Source: Go Facts Geography: *World Events*, pp. 10–3, Blake Education

TARGETING HASS 6 © PASCAL PRESS ISBN: 9781925726077

Research

How has Fukushima and Japan recovered from this disaster? What are some of the long-lasting impacts? Write five interesting facts.

1 ______________________________

2 ______________________________

3 ______________________________

4 ______________________________

5 ______________________________

Questioning

Imagine you were interviewing a person who lived in Fukushima when the earthquake and tsunami struck. What would you ask them about their experiences? Write four questions.

1 ______________________________

2 ______________________________

3 ______________________________

4 ______________________________

Analysing

How do the images accompanying the text contribute to your understanding of this event?

____

Communicating

Create a news report for the day after the earthquake and tsunami.

Think about your format. Will it be a written text, radio broadcast, podcast or video news feature? Present it to the class.

Evaluating and Reflecting

Do you think Julia Gillard was right in saying that Australia shouldn't use nuclear energy? Explain your answer.

____

UNIT 19

Connecting with Asia

In 2019, Australia exported more than $60 billion worth of iron ore to China.

Australia has strong connections with Asia.

After the British established colonies in Australia, many British people migrated to Australia. But when gold was discovered in the 1850s, the mix of nationalities in the country began to change dramatically. The gold rush brought more than 40 000 Chinese people to Australia. Today, one-third of the Australians who were born overseas come from Asia, with the top three Asian countries being China, India and the Philippines. The total number of Australians born in Asian countries represents more than 12% of the Australian population, according to the 2018 census.

Australia's connections with Asia are rarely more obvious than when considering trade. Six of the top ten countries that buy Australian goods and services are Asian. Asia is also where Australians buy almost 60% of their goods and services.

Australia is the top destination for Indonesians wishing to study abroad.

Nearly four million Asians visit Australia each year. They represent more than 42% of all visitors to the country. Most visitors are from China, Japan, Singapore and Malaysia. For Australians travelling to Asia, the most popular destinations are Indonesia, Japan, Singapore, Thailand and China.

Many Asians study at Australian colleges and universities after leaving high school. Their time in Australia benefits everyone at these institutions, as well as the wider community. Everyone gets to experience and expand their knowledge of other cultures and languages. About one-third of students remain in Australia as permanent residents.

Australia has formed defence relationships with several Asian countries. These involve trading military equipment, training together and sharing military intelligence. For example, Australia has a treaty with Indonesia, its closest Asian neighbour, that commits each country to consult on defence, counterterrorism and intelligence sharing. In 1999–2000, Australian troops acted as peacekeepers in Timor-Leste as it progressed from Indonesian control towards independence. Since 2001, Australian troops have also had a role in the West Asian countries of Iraq and Afghanistan. This has included training local forces to defend themselves and rebuilding damaged facilities.

The arts expose people from all over the world to different ideas, experiences and cultures. Whether it is marvelling at ancient terracotta sculptures from China or watching one of India's Bollywood films, these experiences give Australians an insight into the rich history and cultures of Asia.

To promote Australian arts, many organisations travel to Asia to perform and display work. This gives Asian people an insight into Australia's culture, history and peoples. In 2020, the National Gallery of Australia took an exhibition of Aboriginal and Torres Strait Islander art to Singapore and China.

The arts are not always displayed in galleries. In public spaces throughout Australia, audiences can see lion dancers perform and lunar lantern displays light up the skies in celebration of the sounds, colours and cultures of Asia.

A monk makes an offering to the lion dancers at the Chinese Lunar New Year Festival in Hurstville, NSW.

Source: Go Facts Geography: *Asia*, pp. 24–27, Blake Education

GEOGRAPHY

Research

Trade between Australia and Asia is significant. What are the major imported and exported goods and services between Australia and Asia?

Questioning

What are five questions young people should ask about Australia before deciding whether or not to come here to study?

1

2

3

4

5

Analysing

Why do so many Asian students remain as permanent residents in Australia after they have completed their studies?

Communicating

Haiku is a form of Japanese poetry. It is made up of 3 lines that do not rhyme. The first and last lines have 5 syllables and the middle line has 7 syllables.

Create your own haiku poem to share with your class.

Evaluating and Reflecting

Why is it so important for people of different cultures to make connections? Explain your response with examples.

Australia's Aid Program

Australia provides aid to countries around the world to help improve the lives of their citizens. Most of Australia's aid goes to its neighbouring countries, many of which are among the poorest in the world. As well as helping the countries who receive aid, the aid program also helps Australia; it builds connections and helps promote stability in the region.

What form does aid take?

Aid can take many forms, including money, medicines, medical equipment and building materials. It can also be the construction of schools or health clinics, or the provision of clean water, teachers, nurses or doctors. It can even be advice.

Who receives aid?

Australia gives long-term aid to developing countries that need assistance to provide basic goods and services such as food, schools, water, electricity and health care. It also provides emergency aid to help countries that have been affected by war or natural disasters such as tsunamis, cyclones and landslides.

Where is the aid given?

Australian estimated aid allocations per region (2019–20) ($ millions)

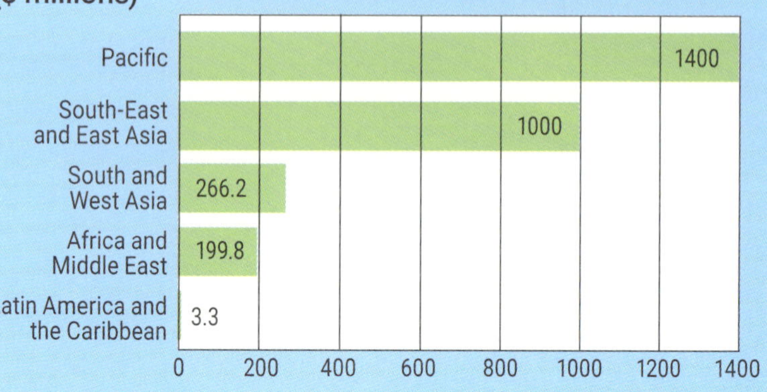

Source: Commonwealth of Australia, DFAT, 2020

CASE STUDY 1 – Papua New Guinea

Papua New Guinea is Australia's closest neighbour. In 2016, Papua New Guinea received around $541 million worth of aid from Australia. This was around 14% of all aid Australia gave to foreign countries. This money has been used to improve transport systems including roads and airports; to train nurses, doctors, teachers and police officers; to build schools and universities; and to provide medical resources including equipment and medicines.

Many people in Papua New Guinea live in small villages.

A lot of infrastructure, such as roads, in Papua New Guinea is very basic.

CASE STUDY 2 – Tropical Cyclone Winston

Tropical Cyclone Winston hit Fiji and Tonga in February 2016. It was the strongest cyclone ever recorded in the southern hemisphere at that time. It killed 44 people and affected two-thirds of the population of Fiji (around 540 000 people). Australia provided $15 million in immediate assistance to supply shelter, water, food, health care and education to those affected. To help with long-term recovery, Australia committed $20 million to rebuild infrastructure such as houses, schools and health clinics.

Tropical Cyclone Winston caused severe flooding in Fiji.

The military is often called upon to help after a natural disaster.

Source: Australian Geography Centres: *Upper Primary*, p. 59, Blake Education

GEOGRAPHY

TARGETING HASS 6 © PASCAL PRESS ISBN: 9781925726077

Research

Which countries in the Pacific region receive foreign aid from Australia? How much do they get? What do these countries use the money for?
Hint: You could start by looking at the Australian Department of Foreign Affairs and Trade website.

Questioning

Write four questions to test a friend's knowledge of the text.

1 ____

2 ____

3 ____

4 ____

Analysing

What does the graph titled 'Where is the aid given?' tell you? Why do you think the money is allocated in this way?

Communicating

Create a piece of word art to represent the topic of Australia's Foreign Aid.

You may choose to use digital tools or complete the task on paper.

Evaluating and Reflecting

Should Australia support other countries financially? Does the Australian government give enough money in foreign aid? Explain your response.

UNIT 21

Australia's Global Connections

Australia might not be physically joined to any other country, but it has connections with many nations throughout the world.

Aid Connections

The Australian Government gives aid to many developing countries around the world. It also provides aid to countries that have just suffered a natural disaster.

Family Connections

Australia is one of the most ethnically diverse nations on Earth. People who live in Australia have families spread out across the planet.

Sister City Connections

Australia has many 'sister city' relationships around the world. These are agreements between cities or towns that promote cultural relationships between the two places. One of Sydney's sister cities is Nagoya, Japan.

Commonwealth Connections

Australia is a member of the Commonwealth of Nations, which is made up of 54 countries or territories. Queen Elizabeth II is the Head of the Commonwealth. Every four years, member countries compete in the Commonwealth Games. Every second year, representatives from each country meet at the Commonwealth Heads of Government Meeting (CHOGM) to discuss issues that affect them.

Organisation Connections

Australia is a member of the United Nations. This is one of several international organisations that Australia belongs to.

Other Connections

Australia has many other connections around the world through humanitarian efforts (refugees), tourism, the environment, religion and trade.

Education Connections

Many university students from around the world come to Australia to study. Most come from Asia.

Sporting Connections

Each year, Australians compete in international sporting events such as the Olympics, the FIFA World Cup and the Netball World Cup.

Scientific Connections

Australia has bases in Antarctica where Australians study issues such as climate change. Australia shares what it learns with many different countries around the world.

Source: Australian Geography Centres: *Upper Primary*, p. 57, Blake Education

Research

Does the place where you live have a sister city somewhere in the world? What is it and how does the relationship work?

If not, then investigate Sydney's sister city relationship with Nagoya, Japan.

Questioning

Write four questions you have after reading the text.

1

2

3

4

Analysing

From your investigation into a 'sister city' relationship, what are the advantages and disadvantages of this type of relationship?

Advantages	Disadvantages

Communicating

Create an imaginative recipe for what it takes to make a 'perfect' world. What ingredients do you need? How much do you need? When and how do you mix them together? How long does it take to cook?

Evaluating and Reflecting

Is it important for an island nation like Australia to have connections with many nations throughout the world? Why? What would life be like in Australia if we didn't have the types of global connections (sporting, trade, education, family) that we have now?

UNIT 22

First Peoples

First peoples are the indigenous peoples of the world. They are descended from the first people to live in their land.

There are more than 370 million indigenous people in the world, belonging to about 5000 people groups, from every continent except Antarctica. They belong to thousands of nations or tribes, many with their own language and customs.

Yagua men in the Amazon preserve their language and customs.

European nations sent ships all over the world between the 1400s and the 1700s. People from countries such as Great Britain, France, Spain and Portugal landed throughout Africa, Oceania, Asia and the Americas. The took the land, food, gold and silver they found, usually by force. They introduced diseases that killed millions of indigenous people. They introduced their own customs and religion.

Latin America is home to more than 400 indigenous people groups. Indigenous peoples have lived there for more than 12 000 years. During the Spanish and Portuguese invasions in the fifteenth and sixteenth centuries, the indigenous population fell from about 150 million to 11 million – mainly due to diseases introduced by Europeans.

Nine out of 10 indigenous people live in only five countries: Bolivia, Guatemala, Peru, Ecuador and Mexico. The majority of indigenous people in Bolivia, Brazil and Chile live in urban areas.

Did you know?

Latin American countries have between seven and 200 languages, except for Uruguay. It has one predominant language – Spanish.

Guarani Indians protesting about the use of their land near São Paulo, Brazil.

Just like first peoples across the world, many native South Americans are battling for survival. They struggle to maintain contact with their land and to benefit from its resources. Armed conflict drives many peoples from their lands. It is hard to keep languages and customs alive. Most of the indigenous population of Peru is poor, and more than half is extremely poor and living in poverty. The World Bank defines poverty as living on $1.90 a day. This means they are often unable to provide basic needs such as food, clothing and education. Some people groups, such as the Tetete of Ecuador, have become extinct.

Some indigenous peoples in Latin America still live traditionally, without contact with the outside world. These groups are known as 'uncontacted'.

The Ayoreo-Totobiegosode people choose to live in traditional ways in the forests of Paraguay. They hunt and farm squashes, beans and melons. For more than 40 years, they have resisted being forced from the forests. Some have died for this resistance, through violence and disease. They remain threatened as forest land is cleared to graze cattle.

Source: Go Facts Geography: *First Peoples*, pp. 4–5, pp. 18–19, Blake Education

GEOGRAPHY

TARGETING HASS 6 © PASCAL PRESS ISBN: 9781925726077

Research

Three well-known Indigenous peoples from Latin-American rainforests are the Shuar, Yanomami and Kogui.

Choose **one** group to investigate.
Write about your most interesting discoveries.

Questioning

Anthropologists study different cultures. Imagine you were an anthropologist. Write four questions you would like answered about the Indigenous group from your research.

1

2

3

4

Analysing

What does the photograph of the Yagua men in the Amazon tell you about their culture and way of life?

Communicating

Create a poster to show others aspects of your own culture.

Evaluating and Reflecting

Should Indigenous groups such as the Ayoreo-Totobiegosode remain 'uncontacted' and continue living their traditional lifestyles? Explain your response.

UNIT 23

Mark's Diary

Did you know?
A third of Laotians live on less than $1 a day. Another third lives on less than $2 a day.

Responding quickly to world events is important. International development workers also help people improve their lives in the long term.

Local villagers built this dam with support from Oxfam. It doubled the yield of their crops.

Sunday: I'm off to visit the Toumlane district in southern Laos. It's about 50 kilometres from the town of Salavan, but the road is dirt most of the way. It will take at least four hours on my bike, so I leave straight after breakfast. I have to wade across the Sedone River, holding my bike over my head. A bridge over the river was bombed in 1973, during the Vietnam War, and there has been no money to repair it since. I meet my workmates at the local guesthouse. There's no glass in the windows and no electricity. We wash in the river nearby.

Monday: We visit the dam the local people built, with money and engineering advice from us. It's made real improvements – farmers now have guaranteed water for one rice crop a year. For a long time, most of the soil here was too poor to grow even one good rice crop. Raising chickens is the other way families have improved their lives.

A water pump delivers clean water where it is needed.

Tuesday: Access to drinking water is a big issue – it's so dry in this district. It's the women's job to collect water, and they sometimes spend all night fetching water! That's the only time it's cool enough for the three-hour walk to the nearest spring. We have a program here to drill deeper wells, right in the villages, and install hand pumps that the women maintain. The water quality is better and the women can sleep at night instead of hauling water!

A group of Lao Theung women

Mark Deasey worked in Laos for Oxfam (www.oxfam.org.au). He currently works for Australian Volunteers International (www.australianvolunteers.com).

Source: Go Facts Geography: *World Events*, pp. 28–9, Blake Education

GEOGRAPHY

Research

Mark Deasey works for Australian Volunteers International (www.australianvolunteers.com). What is their role? Write five facts from your investigation.

Questioning

Imagine you were able to interview Mark about his work in Laos. Write four questions you would ask him.

1

2

3

4

Analysing

What impact do the photographs have as you read the text? How do they help you understand Mark's work?

Communicating

Create a mind map that communicates what you have learnt about organisations that assist people in foreign countries.

Evaluating and Reflecting

Could you live on less than $2 a day, like two-thirds of Laotians? What changes would you need to make to your way of life?

UNIT 24

Cultural Diversity

Culture includes the language, religion, beliefs, traditions, art, music, dance, clothing, shelter and celebrations of different ethnic groups. There are hundreds of different ethnic groups around the world, and it is these groups that make the world culturally diverse.

Cambodia

Location: South-East Asia
Population: 16.2 million
Languages:
Official – Khmer; Other – French, English
Ethnic groups: Khmer 90%, Vietnamese 5%, Chinese 1%, Other 4%
Religions: Majority – Buddhism 96.9%, Other – Islam, Christianity

Traditions and celebrations: Many Cambodian traditions and celebrations reflect Buddhist beliefs. *Pchum Ben* (Ancestors' Day) lasts 15 days. It concentrates on blessing the souls of people who have died. Most Cambodians visit Buddhist temples to make offerings of food to monks who pray for their loved ones' souls and to give food to feed the souls of those who have passed away. Other celebrations reflect the region's farming heritage and dependence upon the environment. The Water Festival celebrates the Mekong River and its gifts of food and water, while Royal Ploughing Day celebrations are conducted to ensure a good harvest.

Angkor Wat in Cambodia is the world's largest religious monument. It was once a Hindu temple, but today is a Buddhist temple and Cambodia's main tourist attraction.

Buddhist monks at Angkor Wat

Cambodian dancers in traditional costume

Boat races are held during the Water Festival.

Egypt

Location: Africa
Population: 98.4 million
Languages:
Official – Arabic; Other – English, French
Ethnic groups: Egyptians 99.6%, Other 0.4%
Religions: Majority – Islam 90%, Other – Christianity, Judaism

Traditions and celebrations: Many Egyptian traditions and celebrations revolve around Islamic beliefs. *Ramadan* is a month-long festival during which Muslims fast from sun-up to sundown. The *adhan* (Islamic call to prayer) is heard five times a day and is a signal for Muslims to stop what they are doing and pray. Egyptian celebrations also reflect the peoples' traditional dependence upon the environment in which they live. Ancient Egyptians relied on the flooding of the Nile to fertilise their fields. *Leylet en Nuktah* celebrates this flooding while *Sham el-Nessim* celebrates the arrival of spring.

The tanoura dance is a traditional Egyptian dance.

A mosque is a building where Islamic people worship.

The Nile River in Cairo has always been important to the Egyptian people.

Many Egyptian women today wear traditional Islamic dress.

Source: Australian Geography Centres: *Upper Primary*, p. 53, Blake Education

GEOGRAPHY

TARGETING HASS 6 © PASCAL PRESS ISBN: 9781925726077

Research

Angkor Wat in Cambodia is the world's largest religious monument. Investigate this human feature and write five interesting facts about it.

1 ______________________________

2 ______________________________

3 ______________________________

4 ______________________________

5 ______________________________

Questioning

Use the text to help you write questions that have these answers.

Answer: Arabic

Question: ______________________________

Answer: Pchum Ben

Question: ______________________________

Answer: Tanoura

Question: ______________________________

Answer: Khmer

Question: ______________________________

Answer: Ramadan

Question: ______________________________

Analysing

How are the cultures of Cambodia and Egypt similar?

____

Communicating

Imagine you were able to visit Angkor Wat in Cambodia. Create a postcard and write a short message to your family in Australia, telling them about your holiday experiences.

Evaluating and Reflecting

Is cultural diversity a positive feature of our society? Would it be better if all people from around the world had the same language, religion, beliefs, traditions, art, music, dance, clothing, customs and celebrations? Give reasons for your response.

____

UNIT 25

Australians as Peacekeepers

The United Nations

United Nations Headquarters, Switzerland

The formation of the United Nations (UN) was started by the Allies of World War II. The United Nations was officially formed on 24 October 1945, when the United Nations Charter came into effect. UN involvement in peacekeeping began in 1948, when UN military observers were sent to help bring stability to the Middle East — a mission that continues to this day. Since then, the UN has conducted over 65 peacekeeping operations throughout the world. Hundreds of thousands of military personnel, police and civilians from more than 120 countries have participated in these operations. Australia is a founding member of the United Nations and has been supportive of its role in human rights, environmental affairs and international economic assistance.

Australia and the UN

Flag of the United Nations

Australia was one of the first 51 original members of the UN. It undertook its first UN peacekeeping responsibilities as military observers in Indonesia in 1947. Since then, Australia has contributed to over 50 peace and security operations around the world. In 1993, the Australian Defence Force (ADF) Peace Operations Training Centre was created to provide more specialised training to Australians sent on peacekeeping missions. Australia has been in command of six multinational peacekeeping operations since 1950. The most recent operation was from 1999 to 2000 in Timor-Leste (East Timor).

Australian Police

Australian police officers have also contributed to UN peacekeeping missions. They have served in Cyprus since 1964, helping to minimise the conflict between the country's Greek and Turkish communities. Australian police have also served in Cambodia, Haiti, Mozambique, Bougainville and Timor-Leste.

Did you know?
Sixteen Australians have died during peacekeeping missions.

Recent Peacekeeping Missions

The Australian-led Regional Assistance Mission to Solomon Islands (RAMSI) began in 2003 with assistance from 15 countries. RAMSI was formed in response to a request for aid from the Solomon Islands Government. At the time, economic problems, violence and corruption were major issues. After much change in the Solomon Islands, RAMSI has now shifted to an ongoing policing mission.

Operation Tower was the name of the ADF contribution to the United Nations Integrated Mission in Timor-Leste (UNMIT). This mission, established in 2006, was the fifth UN mission in East Timor since 1999. Timor-Leste only became an independent nation in 2002, and the aim of UNMIT was to support the security and development of the country. The mission ended in 2017.

Australian peacekeepers are also involved in long-term missions in Sudan under Operation Azure (since 2005), and Operation Hedgerow (since 2007).

Government Palace, Timor-Leste

Source: *Blake's Australian History Guide*, p. 91, Pascal Press

CIVICS AND CITIZENSHIP

TARGETING HASS 6 © PASCAL PRESS ISBN: 9781925726077

Research

Visit the Australian War Memorial website at www.awm.gov.au. Find out more about **one** of the six multi-national peacekeeping operations that have been commanded by Australia. What kinds of responsibilities did the peacekeepers have?

Questioning

If you were able to interview an Australian peacekeeper, what would you ask them? Write four questions about their role and its impact on them.

1

2

3

4

Analysing

Is the Australian War Memorial website you used for your research a primary or secondary source?

What examples of primary sources of evidence did you find on the website? List four.

1

2

3

4

Communicating

Design a commemorative coin, stamp or sculpture to honour and remember servicewomen and servicemen who have been involved in peacekeeping missions. You may choose to use 3D creative software.

Evaluating and Reflecting

How do Australia's peacekeeping missions reflect our role as global citizens? Should Australia as a nation and Australians as individuals get involved in issues happening in other countries? Explain your ideas.

UNIT 26

Australia's Political System

STATE

STATE OR TERRITORY PARLIAMENT
- Makes laws for their state or territory
- Approves laws made by local councils

STATE OR TERRITORY GOVERNMENT
Responsible for areas that are not federal responsibilities (e.g. education, health, environment, transport)

SYSTEM OF PARLIAMENT
The Australian system of parliament is based on the Westminster System. It originated in the United Kingdom in the Westminster district of London, England, where the British Houses of Parliament are located.

STATE GOVERNOR
The monarch's representative, appointed by the monarch on advice from the premier

PREMIER/CHIEF MINISTER
- Leads the government
- Must receive majority support from the political party who wins the election

SYSTEM OF GOVERNMENT
Australia is a constitutional monarchy and a parliamentary democracy. The British monarch is our head of state.

LEGISLATIVE ASSEMBLY/ HOUSE OF ASSEMBLY
Lower House
- Made up of members that represent each electorate
- Government is formed by the party, or coalition of parties, with the majority of members

LEGISLATIVE COUNCIL (NSW, Vic., Tas., SA, WA)
Upper House
- Responsible for monitoring and reviewing laws, legislating, and gathering information
- Holds the government to account

STATE CABINET
- Makes policies for the government
- Made up of the premier and state ministers

LOCAL COUNCIL
- Makes local laws (called by-laws) for its region or district
- Manages public property, services and activities

MAYOR, LORD MAYOR OR SHIRE PRESIDENT
Elected leader of the council

COUNCILLORS
Representatives elected to the council

FEDERAL

AUSTRALIAN PARLIAMENT
- Represents the people of Australia
- Makes and changes laws
- Provides a place where government is formed
- Keeps the government in check

AUSTRALIAN GOVERNMENT
Passes laws that affect the nation (e.g. taxation, defence, foreign affairs)

BRITISH MONARCH
Constitutional monarch, not involved in the Australian government. Holds a symbolic role only

GOVERNOR-GENERAL
- Represents the British monarch in Australia
- Appointed by the monarch on advice from the prime minister
- Independent from the British government

Opening of the first Parliament

PRIME MINISTER
- Head of the Australian government
- Also a member of parliament (MP)
- Leads the parliamentary party or coalition of parties

Senate (Upper House)
Made up of 76 senators who represent Australia's states and territories

House of Representatives (Lower House)
- Made up of 150 members (MPs) who represent Australia's 150 federal electorates
- Whichever party has the majority can form government

CABINET
- Made up of the prime minister and senior ministers
- Role is to direct government policy

Source: *Blake's Australian History Guide*, pp. 72–3, Pascal Press

TARGETING HASS 6 © PASCAL PRESS ISBN: 9781925726077

What are the roles and responsibilities of the different levels of Government in Australia?

UNIT 26

Research

Australia's parliament is based on the Westminster System. What is this?

What role does the governor-general play in Australia's political system?

Questioning

1 What is the leader of a local council called?

2 What is the state legislative council responsible for?

3 What is the role of a governor?

4 What is a cabinet in federal parliament?

5 Who do senators represent?

Analysing

Compare and contrast Australia's **state** and **federal** political systems.

Similarities	Differences

Communicating

Create a piece of word art to demonstrate your understanding of Australia's political system. You may use digital tools or get creative with paper and coloured pens.

Evaluating and Reflecting

In 1975, Australia's Prime Minister, Gough Whitlam, was dismissed from office by the Governor-General, Sir John Kerr. Should a governor-general be able to remove a prime minister? Give reasons to explain your point of view.

UNIT 27

Australia's Legal System

COMMON LAW

The basis of Australia's legal system comes from common law, which developed in the United Kingdom. Common law is a system where judges must base their decisions on the rulings made in previous cases.

FEDERAL

Court	Decided by	
FEDERAL COURTS – Have jurisdiction over laws made by Federal Parliament		
HIGH COURT • Disputes involving the Constitution • Hearing final appeals against decisions made in other courts	Chief Justice	6 Judges
FEDERAL COURT Criminal and civil cases relating to Commonwealth law (e.g. immigration, customs, taxation)	Judge	Jury (very rarely)
FAMILY COURT Family law matters (e.g. parental disputes, child support, custody)	Judge	
FEDERAL CIRCUIT COURT Shares jurisdictions with Federal Court and Family Court (e.g. family law, administrative law, bankruptcy, copyright) and deals with less complicated matters	Judge	

JURY SYSTEM

A jury is a group of citizens sworn in to reach a verdict, based on evidence presented to them. Jurors are selected at random from the electoral roll. There are usually 12 people on a jury.

STATE

Court	Decided by	
STATE AND TERRITORY COURTS – Have jurisdiction over laws made by State Parliament		
SUPREME COURTS COURT OF APPEAL • Appeals regarding decisions made by lower courts	3–5 Judges	
TRIAL DIVISION • Civil matters (e.g. property disputes) • Criminal cases (e.g. murder)	Judge	Jury
DISTRICT AND COUNTY COURTS Serious offences (e.g. armed robbery, assault)	Judge	Jury
LOCAL AND MAGISTRATES COURTS Lesser offences (e.g. traffic violations, burglaries)	Magistrate	

Supreme Court, Perth

Source: *Blake's Australian History Guide*, pp. 74–5, Pascal Press

CIVICS AND CITIZENSHIP

TARGETING HASS 6 © PASCAL PRESS ISBN: 9781925726077

Research

Federal and state courts are one way that laws are made in Australia. Our state and federal parliaments also make laws. Find out how an idea for a law becomes a bill and then goes on to become a law in federal parliament.

Questioning

1 What matters are heard by the Family Court?

2 What is the name of the most powerful court in Australia?

3 How many judges sit on the Supreme Court?

4 Would the District Court listen to a case about the theft of a scooter? Why?

Analysing

Why would the federal courts use judges to listen to the evidence and make decisions, instead of juries made up of ordinary citizens? Hint: Think about the types of cases heard in federal courts.

Communicating

Using your research, create a flow chart to show the steps needed for an idea to become a bill in parliament and then to be passed as a law.

Evaluating and Reflecting

Would you like to be a member of a jury, deciding whether a person is guilty or innocent of a crime? Give reasons to explain your answer.

Being an Australian Citizen

People from every country in the world were represented in Australia's population in 2018. That same year, 534 100 people immigrated to Australia. But not all of these people choose to become Australian citizens.

To become an Australian citizen (if you were not born in Australia), you must:

- have been a permanent resident in Australia for at least one year
- have lived in Australia for at least nine out of the twelve months before you apply for citizenship
- have been a legal resident in Australia for at least four years
- have a good moral character.

The application process to become an Australian citizen generally takes 10–14 months.

When a person becomes an Australian citizen, they are required to take a pledge or oath. Over the years, the wording of this has changed.

2018 Australian Population (24 992 000)		
Birthplace	Number	%
1. England	992 000	4.0
2. China	651 000	2.6
3. India	592 000	2.4
4. New Zealand	568 000	2.3
5. Philippines	278 000	1.1
6. Vietnam	256 000	1.0
7. South Africa	189 000	0.8
8. Italy	187 000	0.7
9. Malaysia	174 000	0.7
10. Scotland	135 000	0.5
All overseas born	7 342 000	29.4
Australian born	17 650 000	70.6

Source: https://www.abs.gov.au/ausstats/abs@.nsf/lookup/3412.0Media%20Release12017-18

1949: Australian nationality created by the *Nationality and Citizenship Act* 1948

I swear by Almighty God that I will be faithful and bear true allegiance to His Majesty King George the Sixth, his heirs and successors according to law, and that I will faithfully observe the laws of Australia and fulfil my duties as an Australian citizen.

1966: The Holt Government added a clause for people to renounce former allegiances
1973: A reference to the Queen of Australia was inserted

I, renouncing all other allegiences, swear by Almighty God that I will be faithful and bear true allegiance to Her Majesty Elizabeth the Second, Queen of Australia, Her heirs and successors according to law, and that I will faithfully observe the laws of Australia and fulfil my duties as an Australian citizen.

1994: Prime Minister Keating replaced the oath with a Pledge of Commitment

From this time forward [under God], I pledge my loyalty to Australia and its people, whose democratic beliefs I share, whose rights and liberties I respect, and whose laws I will uphold and obey.

Source: https://parlinfo.aph.gov.au/parlInfo/search/display/display.w3p;query=Id:%22library/prspub/1VW76%22

When migrants become Australian citizens, they have the same rights and responsibilities as citizens who were born in Australia. They can apply for an Australian passport and vote in Australian government elections. They can also be elected to parliament. Citizens also need to follow the law, serve on a jury if given a summons, and, if necessary, defend Australia.

Research

What does it mean to be a **dual citizen**? How does it impact a person's Australian citizenship if they are a dual citizen?

Questioning

Write four questions that can be answered by a friend using the text.

1

2

3

4

Analysing

How has the citizenship oath or pledge changed over the years? Why has it changed in this way?

Communicating

Create a graph to visually represent the population data in the table in the text.

Evaluating and Reflecting

Should migrants become Australian citizens? Or should they remain citizens of the country of their birth? Give reasons for your answer.

UNIT 29

Trade-offs and Costs

How do you like to spend your weekends? Do you like to go bowling with your friends? Do you enjoy a game of minigolf? Maybe you prefer to play video games or take your dog for a walk?

Trade-offs

It would be nice if we were able to do everything that we wanted. But as soon as we choose one thing to do, it means that we can't do something else. Take homework, for example. If you spend your afternoon doing an hour of maths problems followed by writing an English creative writing story, you won't be able to ride your bike to your friend's house. This is known as a trade-off. It is a decision that involves losing something (the time to ride your bike) in order to gain something else (finishing your homework). It is a balancing act.

Scarcity

The economic reason behind trade-offs is the concept of scarcity. The resources used to produce goods and services are limited (scarce), as is the time we have to enjoy them. But our human needs and wants are limitless. There is greater demand than supply.

Opportunity costs

So, when a person chooses one option over the **next best alternative**, there is a cost involved. This is the cost of missing out on the benefits of the activity, good or service that has been given up. It is called an opportunity cost. Such costs might be explicit – one activity could be more expensive than the other. But opportunity costs might also be implicit – making a new friend by joining a chess club, compared to improving your physical fitness by attending soccer training. Consumers must weigh up the explicit and implicit opportunity costs of their choices. Every choice has a value that is unique to the person making that choice.

When making financial decisions, we try to choose the options which have the greatest benefit to us with the lowest cost.

Real world examples

As a consumer, you have to make decisions about which goods or services to buy and therefore which goods or services not to buy. Do you choose to pay $4 for a salad wrap instead of a chicken roll? If you do, the opportunity cost is not the $4 you spend on the salad wrap, it is the loss of not having the chicken roll.

Businesses also make decisions which involve opportunity costs. A printing company might decide to use non-recycled paper instead of recycled paper. The opportunity cost of that decision would include the environmental cost as well as the financial cost.

TARGETING HASS 6 © PASCAL PRESS ISBN: 9781925726077

Research

Businesses are always making decisions and trying to sell new products, or redesigning existing products to make them appear 'new'.

Sometimes a new product, like the first smart phone or ride share company, changes the market forever. But sometimes the idea fails completely. This happened to Cadbury® when it changed its Roses chocolates range in 2018, and Arnott's® when it changed the flavours of its Shapes biscuit range in 2016.

Choose one of these business decisions and find out what happened.

____

____

____

____

____

____

Questioning

Use the information gathered during your research to answer the following questions.

1 Why did the company change the product?

____

____

2 What **trade-off** did the company make when it changed the product?

____

____

____

____

3 What was the **opportunity cost** of the decision?

____

____

____

____

Analysing

What were the impacts of the decision by the company to change its product? Complete the PMI chart below, based on your research.

Positive	Minus	Interesting

Communicating

Create your own new biscuit or chocolate flavour. What would it be?

Design a magazine advertisement for your product and share it with your friends. Do they want to purchase it? Discuss why.

Evaluating and Reflecting

Did the company you researched make the right decision in changing its product? What else could they have done? Give reasons for your response.

____

____

____

____

Making a Profit

What is a business?

Supermarkets, milk bars, service stations and newsagencies are all businesses. But what about the people who mow lawns for a living, or a teacher who tutors privately on the weekend? Do these qualify as businesses too? We can all identify and name examples of well-known businesses, like McDonalds® and the Commonwealth Bank. But what exactly is a business?

A business is generally understood to be an organisation, entity or activity designed to produce goods or services. Your school is an organisation, but is it a business? Your local football or netball club is an organisation, but is it a business? How is your local club different to an AFL, NRL or A-League club?

A business is an organisation that brings together physical, natural, human and financial resources in order to produce goods or services for consumers, with the aim of making a **profit**. In order to make a profit, the business must have greater revenue (money coming in to the business) than expenses (money going out of the business). Businesses can be run by one individual, like a home-based beautician, or they can be a large multinational company that employs thousands of people around the world.

Profit

Most companies want to make a profit. This is the money they have left over from selling their goods or services and after all their expenses such as wages, taxes, rent and utilities. Any profit a company makes can be:

- invested in the business by purchasing new equipment, technology, additional advertising or resources (retained profits)
- paid to the owners of the business
- saved in cash reserves for the future.

Not for profit

Not all businesses operate to make a profit. Some businesses are not-for-profit. But this doesn't mean they don't want to make money from the sale of goods or services. It means that their profits must be used to help the organisation's mission or goal and the greater community, rather than the personal gain of the business owners.

Organisations such as Care Australia, the Red Cross, The Fred Hollows Foundation and Koorie Heritage Trust are not-for-profit businesses. They have to follow different laws and regulations than for-profit businesses. Not-for-profit organisations must be registered in the state they operate in. They must have an official contact point, hold annual meetings and submit annual reports. Many not-for-profit organisations, such as registered charities, do not need to pay company tax in Australia. Donations to charities and not-for-profit organisations can also be tax deductible.

Source: *Exploring Business Book 1*, p. 9, Blake Education

TARGETING HASS 6 © PASCAL PRESS ISBN: 9781925726077

Research

Select one of the not-for-profit organisations listed in the text. Find out what they do and how they work towards a community-based goal.

Questioning

What are the meanings of these words, as they are used in the text?

- retained
- tax
- utilities
- deductible
- revenue
- expense

Analysing

Research shows that 79% of Australians believe that younger Australians are less actively committed to long-term activities than previous generations. If this is true, how will it impact not-for-profit organisations in the future?

Communicating

Create a short 1-minute pitch to convince someone to donate to a specific not-for-profit organisation.

Present it to your class or a small group.

Evaluating and Reflecting

Should not-for-profit organisations be exempt from paying company tax in Australia? Give reasons for your opinion.

UNIT 31

Printer Cartridge Recycling

By making products from old and used products, we are acting sustainably. We are reducing waste and saving materials. Making products from recycled materials also uses less water and energy than making them from new materials.

Used printer cartridges are one of the fastest-growing forms of waste in Australia. By collecting cartridges and sending them to be recycled, we can recover materials like plastics, steel, aluminium and inks. These can then be used to make brand new products, such as rulers, pens and even the tarmac on our roads.

There are a number of recycling programs that offer Australians a way to recycle their used printer cartridges. It's up to all Australians to limit what we call 'complex waste' products going to landfill. Complex waste includes things with multiple parts. These parts can't be broken down naturally and take up a lot of space. If they weren't recycled, they would end up in overcrowded rubbish dumps. Most of our complex waste comes from computers, printers, televisions, smartphones and other electronic products.

Cartridges 4 Planet Ark

Planet Ark is an Australian not-for-profit organisation that promotes sustainable resource use. They have a printer cartridge recycling program called Cartridges 4 Planet Ark. You can drop off your used cartridges at collection points, or register for your own school to get a free collection box. Each year, Planet Ark collects thousands of tonnes of complex waste and sends it for reprocessing to help the environment.

Source: Australian Geography Centres: *Middle Primary*, p. 59, Blake Education

ECONOMICS AND BUSINESS

TARGETING HASS 6 © PASCAL PRESS ISBN: 9781925726077

Research

Investigate the Cartridges 4 Planet Ark program. Find out five interesting facts about it.

1 ______________________________

2 ______________________________

3 ______________________________

4 ______________________________

5 ______________________________

Questioning

Write four questions you could ask people to survey them about their recycling habits.

1 ______________________________

2 ______________________________

3 ______________________________

4 ______________________________

Analysing

What impact does the image of computer waste have on the reader? Why was it included in the text?

Communicating

Create your own recycling box at school. Design a sign or poster to inform students and teachers about its use.

Evaluating and Reflecting

Should more be done to recycle computer and technological waste? What are some of the financial (not environmental) issues that impact recycling?

Global Financial Crisis

Modern finance is global. Financial trouble in one part of the world can have disastrous effects in another.

An economic recession is a slowing of economic activity. Income and spending falls. Unemployment and poverty rises. The worst economic recession in the past century was the Great Depression of the 1930s. However, other recessions in recent times have been caused by the COVID-19 pandemic in 2020 and the Global Financial Crisis (GFC) in 2007.

The GFC

The GFC began when banks and finance companies in the United States willingly lent money for houses to people considered to be 'investment risks' – people who may not be able to pay the money back.

Interest rates rose and many people failed to make their payments. But then house prices fell dramatically in 2006. The banks could not even sell the houses to get back their money.

US banks took back many homes, forcing people to find new places to live. They attempted to get their money back by selling the properties.

Credit crunch

Governments often help banks when they get into trouble, but in 2008 the US government allowed a global bank, Lehman Brothers, to become bankrupt. All banks were suddenly at risk of failing. Governments began giving them billions of dollars of taxpayers' money. A lack of confidence spread through financial markets. Banks wouldn't lend money (credit) to each other, and this is known as a 'credit crunch'.

The United States has the largest economy in the world, so the lack of confidence spread globally. Companies stopped investing and employing people because they were worried people could no longer buy their goods and services.

Australia was not as badly affected by the GFC as some other countries. First, the Australian banking system was stronger. Second, a lot of Australia's trade was with China, which was not greatly affected by the GFC. Third, the federal government spent a lot of public money, more than $50 billion, to keep people working by helping small businesses, repairing roads and schools and encouraging people to buy products.

Economics is about real people, not just numbers on a computer screen. The COVID-19 crisis affected people's ability to work, support their families and enjoy life.

When the 2020 COVID-19 influenza pandemic struck, it dramatically impacted global trade. The virus epicentre, the city of Wuhan in the province of Hubei, China, was shut down. This completely stopped the export of many products such as electronics, textiles and car manufacturing components.

As the virus spread around the world, countries quarantined themselves, banning travel. Almost overnight, international tourism came to a complete halt. In Australia, the government closed its borders to foreign travellers. QANTAS airlines laid off 30 000 employees and businesses such as gyms, movie cinemas and play centres were closed. Many other businesses encouraged their employees to work from home or take shorter shifts.

The federal government stepped in to offer financial aid to those people who were out of work due to the pandemic. They increased welfare support payments and introduced the JobKeeper payment so that businesses could keep employing their workers. The aim was for those workers to return to their normal job once the crisis was over. These stimulus measures cost the Australian government more than $215 billion.

Source: Go Facts Geography: *World Events*, pp. 28–9, Blake Education

TARGETING HASS 6 © PASCAL PRESS ISBN: 9781925726077

Research

Find out five interesting facts about the federal government stimulus measures to support workers and businesses during the COVID-19 pandemic.

1 ______________________________

____

2 ______________________________

____

3 ______________________________

____

4 ______________________________

____

5 ______________________________

____

Questioning

Write four questions to test a friend's understanding of the text. Use **what, who, where** and **why** in your questions.

1 ______________________________

____

____

2 ______________________________

____

____

3 ______________________________

____

____

4 ______________________________

____

____

Analysing

What would have been the trade-off for the government's decision to spend $215 billion on the COVID-19 stimulus measures? Was the stimulus spending a good decision? Why?

____

____

____

____

____

Communicating

Create a Venn diagram to compare and contrast the GFC with the financial crisis caused by the COVID-19 pandemic.

Evaluating and Reflecting

Research after the Global Financial Crisis showed that most Australians saved the cash they were given, rather than spending it. Was this a good idea? Was it what the government wanted? Explain your answer.

____

____

____

____

____

History Assessment 1

Why and how did Australia become a nation?

Task: Answer these questions.

1 When did Australia become a federated nation? __________

2 Identify two people who were important in the creation of a federated Australia. Briefly describe their roles.

Name:	Name:
Role in Federation	Role in Federation

3 Identify the main reasons for and against Australia becoming a federated nation.

Reasons to support federation	Reasons to oppose federation

4 What might have happened during the following if Australia had not been federated?

Australia's involvement in World War I	
Establishing Aboriginal citizenship rights	
The development of 6 states and 2 territories	

TARGETING HASS 6 © PASCAL PRESS ISBN: 9781925726077

History Assessment 2

How did Australian society change throughout the twentieth century?

When Captain Arthur Phillip raised the British flag on the shores of Sydney Harbour in 1788, it signalled the beginning of great changes in an ancient land.

Over the decades that followed, British convicts and settlers cleared the land and grew crops. They built towns and cities. They built roads and rail lines. They cemented their way of life in the land and created a new piece of the British Empire.

Task: Explain the impact of these things in regard to Aboriginal and Torres Strait Islander peoples' way of life, and their cultural and spiritual link to Country during the twentieth century.

Sewing class at Toomelah Aboriginal Mission, 1934

The Australian Constitution	Aboriginal Missions

State Aboriginal Protectors	The 1967 Referendum

Source: Great Speeches – Saying Sorry – *Speeches about Aboriginal Reconciliation*, p. 2, Blake Education

ASSESSMENT

History Assessment 3

Who were the people who came to Australia? Why did they come?

Task: Use the text to help you answer the questions.

1 Is the image a primary or secondary source?

2 When was the image created?

3 Why was the photograph taken?

4 What information does the photograph give you about refugee camps in Europe after World War II?

5 Is the source reliable? Why?

6 What were the push factors for Marija's migration?

7 What was the pull factor for Marija's migration?

8 What is a refugee?

9 How does Marija feel about migrating to Australia?

Displaced persons in a refugee camp in Europe, 1947

After World War II, many people from Europe could not return to their homelands. They became known as **displaced persons** and lived in refugee camps across Europe. Australia agreed to accept some of these refugees and settle them in Australia.

Hello, my name is Marija and I am seven years old. Before the war, my family and I lived in Latvia. It is a country in Eastern Europe. During the war, life was very dangerous. My family and I left Latvia with only the clothes on our backs. We walked for many days and nights to a refugee camp. After the war we wanted to go home, but the Russians had taken control of Latvia and they treated the people very badly. Now we are afraid to go home. Australia has agreed to take my family. I am frightened because I do not know much about Australia. My dad is the only person in my family who speaks a little bit of English. I hope I will be safe there and that people will be kind to me.

Source: Australian History Centres: *Upper Primary*, p. 63, Blake Education

Geography Assessment 1

How do places, people and cultures differ across the world?
What are Australia's global connections between people and places?

Tasks:

1 On the map, create a colour key to identify the listed countries. The first one has been done for you.

2 Then find out the population and the capital city for each country.

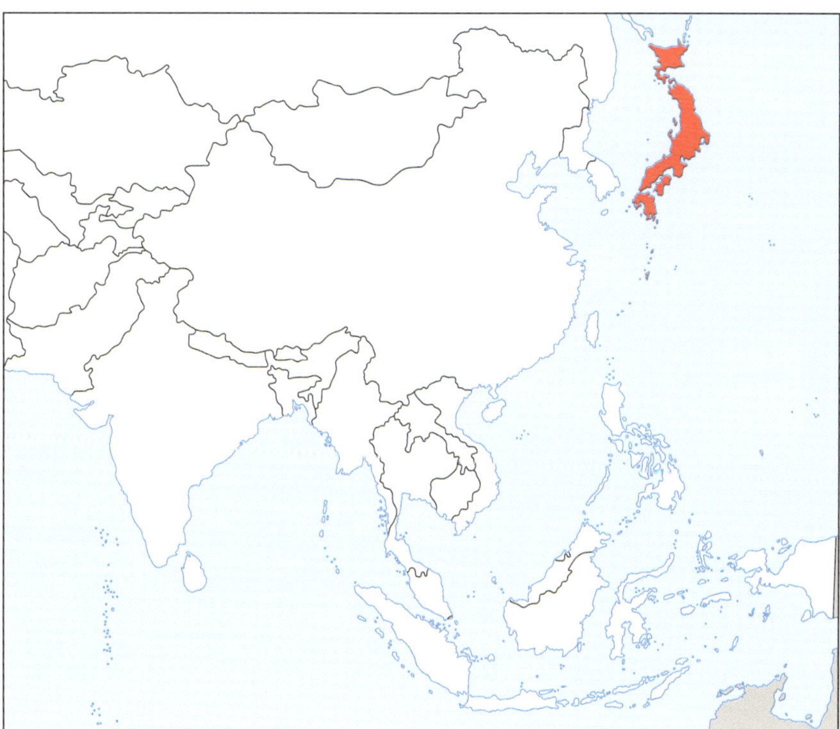

Key

Country		Population	Capital City
	Japan		
	Thailand		
	India		
	China		
	Philippines		
	Malaysia		
	Vietnam		
	Indonesia		

3 Choose one country from the table. How does Australia interact with that country?

____

____

____

Geography Assessment 2

How do places, people and cultures differ across the world?

Indigenous cultures around the world have similarities and differences. Often these are the result of the environment in which they live and the resources they have access to. For example, the Inuit peoples who live in the frozen Arctic region eat very few fruit and vegetables, as these do not grow in a cold environment. One similarity is that today many indigenous peoples around the world live a semi-traditional life – their lives are a mix of traditional and modern ways.

Iban People (also known as Sea Dayaks)

The Iban people live on the island of Borneo and were originally headhunters. Their population numbers around 750 000 and their traditional language is Iban. They live in the hills and plains of tropical rainforests.

Extended families – including parents, grandparents and children – traditionally live in homes called *rumah panjai* (longhouses), which are built along rivers and creeks. Today, many longhouses have modern facilities, such as running water and the internet.

Traditional clothing is usually worn only at special events, such as weddings. Women wear silver bracelets and beads, and decorate their clothing with bells. Men wear a loincloth made from bark. Both men and women have tattoos. Tattoos signify important events in a person's life, and Iban people believe they protect the bearer.

Traditional dances are performed using instruments including gongs and drums. Each dance has a purpose, such as welcoming a visitor or preparing a warrior for battle.

Iban people grow rice – it is their main food, accompanied by jungle fruits and vegetables. They also eat fish, chicken and pork.

Iban people are known for their cloth, basket and mat weaving. Their traditional cloth, called *pua kumbu*, is often used to decorate their homes. Men also make wooden carvings.

1 What continent do the Iban people live in?

____

2 What climatic zone is Borneo in?

____

3 Why would the Iban build houses near rivers and creeks?

____

____

4 What materials would the Iban have access to for building houses?

____

____

5 What weather conditions would the Iban people experience?

____

____

6 How would the climate and environment influence the Iban people's traditional way of life?

____

____

____

____

____

7 List three ways in which the Iban's traditional way of life is different to yours.

____

____

____

Source: Australian Geography Centres: *Upper Primary*, p. 49, Blake Education

TARGETING HASS 6 © PASCAL PRESS ISBN: 9781925726077

Geography Assessment 3

What are Australia's global connections between people and places? How do places, people and cultures differ across the world?

When people study countries around the world, they classify them as **developed** or **developing** countries. Development refers to how advanced a country is, socially and economically.

Developed country: Australia (2019)

Population 25.5 million
Population density 3.3 people per km^2
Life expectancy 85 years (women), 81 years (men)
Per capita income (PCI) $53 825 (USD)

Developing country: Thailand (2019)

Population 69.8 million
Population density 136 people per km^2
Life expectancy 79 years (women), 72 years (men)
Per capita income (PCI) $5650 (USD)

Sources: World Health Organization and The World Bank

1 Write T (true) or F (false) in the boxes to identify the factors that classify a country as developed or developing.

- ☐ A developed country has strong industries such as manufacturing and banking.
- ☐ Developing nations have access to good health care and education.
- ☐ Developing nations have populations that are low and do not grow quickly.
- ☐ The average income in developed nations is high.
- ☐ Developing nations have high population densities.
- ☐ Developing nations have higher life expectancies than developed nations.
- ☐ Developed nations have low unemployment.
- ☐ Stable government is more common in developing nations.

2 What connections does Australia have with Thailand? You will need to do some additional research.

Trade: ______________________________

____

Tourism: ______________________________

____

Defence: ______________________________

____

Culture: ______________________________

____

Source: Australian Geography Centres: *Upper Primary*, p. 37, Blake Education

Civics and Citizenship Assessment

What are the roles and responsibilities of the different levels of government in Australia? How are laws developed in Australia? What does it mean to be an Australian citizen?

1 What is the role of the Senate? ______

2 In which house does the government need a majority? ______

3 Australia is known as a ______ democracy.

4 What is the leader of a state government called? ______

5 What is the role of a local council? ______

______

6 List **three** areas that the federal government is responsible for.

______

7 List **three** areas that the state government is responsible for.

______

8 Who is Australia's head of state? ______

9 Use the numbers 1–10 to sequence these steps explaining how a bill is passed through federal parliament.

- ☐ Once approved by the lower house, the bill goes to the Senate for the first reading.
- ☐ The bill is introduced into the lower house (1st reading).
- ☐ The bill then goes to a Legislation Committee Public Hearing.
- ☐ The Senate considers the bill at the second reading and debates the matter.
- ☐ A bill (proposed law) is drafted by lawyers in the Office of Parliamentary Counsel.
- ☐ The lower house considers the bill for changes and amendments.
- ☐ The bill is read for a second time and debated by parliament.
- ☐ After the third reading, the bill is voted on by the Senate.
- ☐ The bill finally passes to the Governor-General for Royal Assent.
- ☐ At the third reading, the members of the House of Representatives vote on the bill.

10 What does it mean to be an Australian citizen? ______

______

11 What responsibilities do Australian citizens have? ______

______

12 What entitlements do Australian citizens have? ______

______

TARGETING HASS 6 © PASCAL PRESS ISBN: 9781925726077

Economics and Business Assessment

Why are there trade-offs associated with making decisions?
What are the possible effects of my consumer and financial choices?

Mikel had $15 to spend. He really wanted to try out the new $7 lime and berry flavoured slushie drink. But he also needed to replace the grips on the handlebars of his bike, which cost $10. He decided to buy the grips because his mum was always reminding him about riding safely.

1 What trade-off did Mikel make? ______________________________

2 What was the opportunity cost of Mikel's decision?

____

3 Do you think Mikel made a smart choice as a consumer? Why?

____

____

4 Write the following resources in the category that best describes them.

- 3D printer
- coal
- computer
- crane
- filing cabinet
- hammer
- iron
- leadership
- natural gas
- packing boxes
- raw wool
- recruitment services
- safety officer
- sand
- water
- welding machine
- wood
- worker training

Human Resources	Capital Resources	Natural Resources

5 Write a definition for each term, in your own words.

Scarcity: ______________________________

Profit: ______________________________

Business: ______________________________

6 What are three examples of an expense that a newsagent might have?

____

Answers

Note: Answers are not supplied to open-ended questions, where students' responses will vary.

ANSWERS

1 To Federate or Not to Federate?

Questioning: Sample responses would include why they wanted or didn't want change, what were the benefits of their decision, and how they think their decision would impact other Australians.

Analysing: The cartoon is a primary source. It is in favour of Federation. It shows people separated by divided states in the top part. In the bottom part, people are standing together, joined in a common goal as one Australian nation.

2 Timeline Towards Federation

Research: Sample responses:

It was given at the Tenterfield School of Arts. It was the first direct appeal to the general public. People wanted a peaceful transition, unlike the American War of Independence. Parkes had been elected to parliament by the people of Tenterfield. Tenterfield was also important because it was impacted by state tariffs (taxes) between Qld and NSW. Parkes wanted a single national army, not state militias.

Questioning: Responses should include 'when' questions for specific dates of events, 'who' questions for the key historical figures involved, and 'what' questions for the actions that were taken or suggested.

Evaluating and reflecting: 'Yes' responses may cite historical reasons and a sense of continuity. 'No' responses may cite the different views of Aboriginal and Torres Strait Islander peoples.

3 Important Federation Australians

Research: Melbourne and Sydney were both large cities and couldn't agree on which one should be the nation's capital. To reach a compromise, they decided to create a new capital – Canberra. In the Constitution, land was provided from NSW to form the Australian Capital Territory. There is also some belief that it was built in its location because of the cool 'bracing' climate and 'pure' air.

Analysing: Parkes wanted to create a federated nation through peaceful methods. He wanted to break down some of the rights the states had to set tariffs. Barton differed to Parkes in that he wanted the states to keep many of their rights, although he was in favour of free trade. Deakin wanted the Senate, and therefore the power of the states, to be weak. He wanted to limit the Senate's power over financial matters.

4 The Australian Constitution

Research:

Great Britain's constitution is not codified (one single written document). Its constitution is formed from Acts of Parliament, court judgements and conventions. Britain's constitution is abstract and has evolved over a long period of time. The key is the Bill of Rights from 1689 which established that Parliament had more power than the Crown.

The United States of America's constitution is the most important law in the country. It separates the powers of the government into three branches: legislative (congress and the senate); executive (president); and judicial (supreme court and federal courts). The constitution came into force in 1789. It has been amended (changed) 27 times. It asserts that the government is meant to serve the people.

Questioning:

jury: a group of people (usually 12) sworn in to give a decision (verdict) in a legal case in a court of law

constitution: a group of fundamental principles (beliefs) or established precedents according to which a state, country or organisation must be governed.

delegate: a person who has the authorisation to represent others; someone elected to represent others

Act: a bill that has been approved by both houses of parliament

Analysing: The table shows that the three constitutions are very similar, but Australia is more closely aligned to Great Britain than to the United States because of our link to the Queen rather than a President. In terms of levels of government, we are like the USA with federal, state and local government. Great Britain is made up of several countries (England, Ireland, Scotland and Wales) so it doesn't have separate states.

5 Indigenous Rights

Research: The Tent Embassy began with four men with a beach umbrella protesting Prime Minister McMahon's approach to Indigenous land rights. His government had just rejected granting independent ownership of traditional land to Indigenous people, preferring to issue 50-year leases. The Tent Embassy became permanent in 1992 and widened its goals to include Indigenous self-determination. The Tent Embassy represented Aboriginal people's feelings of being alien (foreign) in their own land.

Questioning:

1. Aboriginal people did not receive the same rights and freedoms as other Australians because for a long time, they had not been included as Australian citizens in the Constitution. They were also not included in the census. The federal government did not have the power to make laws for Aboriginal people, only the states did.
2. The states controlled the lives of Aboriginal people through their state protection boards.
3. The 1948 *Nationality and Citizenship Act* meant that anyone born in Australia (including Aboriginal and Torres Strait Islander peoples) was an Australian citizen, which also made them a British subject.
4. The Constitution was amended in 1967 to include Indigenous people in the census and to allow the federal parliament to pass laws for Indigenous peoples.

Analysing: The use of this image creates a sense of compassion for Aboriginal people. But it also could be seen as presenting the Aboriginal peoples as childlike and in need of assistance.

Communicating: Timeline of Aboriginal voting rights

1901: Australian Federation and the adoption of the Australian constitution where Aboriginal people were not counted in the census

1948: *Nationality and Citizenship Act* – all people born in Australia recognised as Australian citizens

Note: Answers are not supplied to open-ended questions, where students' responses will vary.

1962: Amendments to *Commonwealth Electoral Act*. Indigenous Australians could enrol to vote in federal elections, but it was not compulsory.

1967: Referendum to change the constitution to allow the federal government to pass laws relating to Aboriginal people and to include Aboriginal people in the census

1971: Neville Bonner became the first Indigenous Australian to be elected to Federal Parliament

1972: Aboriginal Tent Embassy set up

1979: Australian Electoral Commission began the Aboriginal Electoral Education Program

1990: Aboriginal and Torres Strait Islander Commission founded

1992: Aboriginal Tent Embassy becomes permanent

6 Labourers and the White Australia Policy

Research: Sample responses:

It was known as the White Australia Policy. It impacted migrants from 1901 until 1958. It was created because of concerns about Chinese and Pacific Islander migration. It was one of the first laws passed by the new federal parliament. It was based on existing state laws. Migrants were asked to take a dictation test. They needed to write 50 words in a European language (English, French, Italian, etc.). Migrants could be rejected if they came from an 'undesirable' country, had medical issues or were thought to be 'morally unfit'.

Questioning:

1. False, 2. True, 3. True, 4. True, 5. True, 6. False

Analysing: The cartoon shows the negative attitude by having the sign in the shop window 'Cats and rats wanted'. It also shows how the Chinese diggers bought items that other diggers couldn't afford because they didn't have enough money.

7 Strikes and Suffrage

Research: Sample response:

Henrietta Dugdale was president of the first Victorian Women's Suffrage Society, an organisation she helped form. This model was followed by women in other colonies. She wanted women to achieve equal social, legal and political status with men. Henrietta published a booklet called 'A few hours in a far-off age'. She also attacked the Victorian court system for not taking action on violence against women. She is credited as one of the women who led Australia to become the second country to give women the right to vote.

Analysing: Sample responses:

Role of women in the 1800s	Role of women in modern society
To be in the home To attend to housework and domestic duties To provide children for their husband To be subordinate Not able to vote	To have jobs of their own Able to be in roles of authority in business Able to vote Not expected to do all of the housework To speak their own mind and have their own opinions

Evaluating and reflecting: Sample response:

Advantages: Helped to give women a voice, helped to give women the right to vote, helped women in other countries rally support and make change, allowed greater equality in Australian society.

Disadvantages: There was some violence in clashes between supporters and non-supporters of the suffrage movement, some women were arrested for protesting, our society still does not have equality between men and women.

8 Spanish Influenza

Research: A pandemic is an outbreak of a disease that has spread over a large region or between several countries. Its spread cannot be easily controlled. Pandemics occur at infrequent intervals but there have been several significant pandemics over the past few hundred years. Previous pandemics have been COVID-19 (2020), HIV/AIDS (2005-2012), Hong Kong Flu (1968), Asian Flu (1956-1958), Spanish Flu (1918-1920), Sixth Cholera Pandemic (1910-1911), Russian Flu (1889-1890), Third Cholera Pandemic (1852-1860), The Black Death – also known as the Bubonic Plague (1346-1353).

Questioning:

1. The virus is called 'influenza' because people in Italy believed that the stars 'influenced' diseases and infections.

2. The 1919 Spanish influenza spread around the world because it was carried by soldiers who were returning to their home countries after World War I.

3. The spread of the influenza virus can be slowed by having good personal hygiene, washing hands regularly with soap, sneezing into elbows or a tissue and not hands, and by keeping apart from other people (social distancing).

Analysing:

Similarities	Differences
Both pandemics were types of influenza. Countries put quarantine measures into place. States in Australia closed their borders. Some goods were in short supply (flour in 1919 and toilet paper in 2020).	Higher death toll in 1919. More people infected in 1919. Spread around the world faster in 2020.

9 Reasons for Migration in the 20th Century

Research: The Snowy Mountains Hydro Scheme took 25 years to build and was completed in 1974 at a cost of $820 million. It was built to provide electricity and water to households and for farming irrigation. It collects snow and rainfall from the Snowy Mountains and stores it in dams. The water is then diverted to power stations that generate electricity. It began in 1949 and Australia had to encourage skilled migrants and people displaced after World War II to come and work on the project. It was the beginning of Australia's multiculturalism and was one of the largest construction projects in the world at the time.

Answers

Note: Answers are not supplied to open-ended questions, where students' responses will vary.

ANSWERS

Analysing:

Person already living in Australia	Non-European migrant	British migrant
Support the policy because new migrants might take their jobs and because prices might go up if there are more people wanting to buy goods and services. Support the policy due to fear of people with different beliefs and ways of life.	Be upset by the policy as it discriminates against them. Not understand why they can't be treated equally. Upset by the fact that the policy suggests they are less advanced morally and intellectually than 'white' Australians.	Be supportive because it would give them a better chance of migrating to Australia. Be supportive because the country they were migrating to (Australia) would continue to be similar to their homeland.

Evaluating and reflecting: 'Yes' responses may argue that the policy was good because it kept the community more homogenous. Everyone spoke the same language and had similar values and beliefs and ways of life so there was a greater chance of harmony between migrants and existing Australian residents.

'No' responses may argue the policy was bad because it discriminated against people simply due to the country they were born in. It meant that people who had skills that were needed in a developing nation were excluded from making a positive contribution.

10 Vietnam War and Refugees

Research: After the Vietnamese city of Saigon fell to communist forces in 1975, many Vietnamese people fled to Australia. Tan Thanh Lu, with three friends, built a small fishing boat they called *Tu Do*, which means 'freedom'. Tan used it for fishing before their actual escape to avoid suspicion. In 1977, with 38 passengers (mostly friends, neighbours and relatives) on board, Tan set off. They left at night and had to push the boat for several kilometres through shallow waters so that the noise of the motors did not give them away. They escaped from pirates, and finally landed in Mersing, where eight of the passengers disembarked and were granted refugee status. The others continued on to Australia, landing near Darwin on 21 November 1977.

Questioning: Questions may include asking about his experiences escaping the pirates, what they ate and drank while on the boat, why it was so important for them to leave Vietnam, and what they expected when they arrived in Australia.

Analysing: Sample response:

Fitting so many people in such a small space, personal hygiene issues of washing and toileting, the lack of shade for older people and children in particular, having and storing enough food to support the passengers, how they handled rough weather when waves would have splashed over the sides of such a shallow boat.

Communicating:

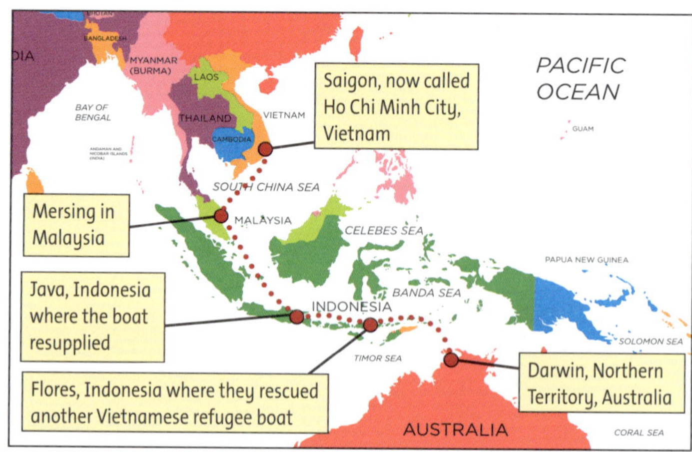

11 Notable Australians

Research: Sample responses:

May Gibbs: (1877-1969) May Gibbs is one of Australia's most famous artists, illustrators and children's authors. Her works created a fantasy world using uniquely Australia wildlife, particularly its plants. She was born in England and migrated to Australia in 1881. She returned to England in 1900 to study art. Her first illustrations were published under the pseudonym 'Blob'. She is most famous for her *Gumnut Babies, Gum Blossom Babies* and of course *The Tales of Snugglepot and Cuddlepie*, which has never been out of print.

Albert Namatjira: (1902-1959) is possibly one of Australia's best-known Aboriginal artists. He painted western-style landscapes. His fame led to him and his wife becoming the first Aboriginal people to be granted Australian citizenship in 1957. As a child, he lived in the Aboriginal mission at Hermannsburg in the Northern Territory. At the age of 13, Albert underwent initiation where he lived in the bush for six months and was taught traditional laws and ways of life by tribal elders. As an adult, he worked as a camel driver, travelling the country that he would later paint.

John Flynn: (1880–1951) After graduating high school, John became a teacher. In 1903 he joined the ministry and was ordained in the Presbyterian Church in 1911. He had a keen interest in working in the outback of Australia. He saw firsthand the hardships of rural life with no medical care. In 1927, Flynn began the Aerial Medical Service and signed an agreement with QANTAS to operate an aerial ambulance from Queensland. In its first year, the AMS flew 50 flights and treated 225 patients. It changed its name to the Flying Doctor Service in 1942 and became the Royal Flying Doctor Service in 1955.

Analysing: They are similar in that they all worked for the greater good of society. They were not promoting themselves as individuals. They each had the goal of improving Australian society and helping those who needed it.

TARGETING HASS 6 © PASCAL PRESS ISBN: 9781925726077

Answers

Note: Answers are not supplied to open-ended questions, where students' responses will vary.

12 A Diverse Land

Research: Possible answers include:

Chinese – food, Chinese New Year celebrations

Italians – food, love of sport such as soccer, many went into farming and created market gardens

Greeks – food, lamb, created milk bars in Australia

Germans – food, introduced Oktoberfest, helped create Australian wine industry

Questioning: Sample responses:

The natural environment: What does high levels of immigration put pressure on?

Soccer: Which game did post-war migrants bring to Australia?

Mosque: Where do many Turkish people practise their religion?

The economy: What is the financial part of a country known as?

Analysing: The overall population trend has been increasing. The natural increase very closely matches the birthrate data. Recently, there has been a more significant increase in the birthrate, probably due to good economic conditions. The rates of migration have been much more varied. The low points in migration followed economic recessions. The migration increases have corresponded with times when Australia needed more skilled migrants. The recent increase is possibly due to the increasing numbers of foreign students living in Australia and studying at universities here.

13 Asia

Research: Sample answers:

China's growing population is putting pressure on the environment through overgrazing of its farmlands and increased levels of logging for timber, leading to increased desertification. There is a higher demand for energy with means increasing use of coal and a related increase in air pollution. There are also the impacts of mining and the release of wastewater.

India's growing population has led to groundwater and surface water contamination. There is increased air pollution from factories and more use of transportation to move products around the country. Lower levels of rain have resulted in lower crop yields. Framers are therefore using more chemicals to try and improve production, but this kills the microorganisms that keep the soil healthy. Forests have been cut down to provide more farming land, resulting in loss of habitat for native plants and animals and lowering biodiversity. Wood is cheaper than gasoline, so farmers burn it for fuel, adding to the levels of pollution in the atmosphere.

Questioning: Sample answers:

In which hemisphere is most of Asia found?

How many countries are in Asia?

How many people live in Asia?

How much of the Earth's land mass is covered by Asia?

Which two other continents share a land border with Asia?

Analysing: Sample answer:

Because Asia is so big, it covers a wide range of environments from tropical rainforests to snow-capped mountains. It contains some of the wettest, driest, hottest and coldest places on Earth, from the equator to the Arctic circle.

14 Mongolia and Pakistan

Research:

Mongolia – Ulaanbaatar 47.8864 degrees N, 106.9057 degrees E

Pakistan – Islamabad 33.6844 degrees N, 73.0479 degrees E

Distance between them: 5 828.8 km via road, 3229.95 km via plane.

Questioning: Sample questions:

What are the official languages of Pakistan? How many people live in Mongolia? What is the major religion of Mongolia? Which climatic zone is Pakistan in? Where do most people live in Mongolia? What is an advantage of the Mongolian *gers*?

Analysing:

Mongolia: small population, Buddhist religion, very cold winters, lakes and grasslands

In common for both: many people live in cities, extreme temperature range, high mountains

Pakistan: large population, Islamic religion, very hot summers, mostly dry, arid and rocky

Evaluating and reflecting: Responses could refer to overcrowding in the cities, not being able to produce enough food for the population, building houses on fertile land which means there is less land for farming, increasing levels of water pollution.

15 Environments

Research: Sample answer:

Animals in Indonesian rainforests include the Sumatran tiger, Komodo dragon, Sumatran orangutan, sun bear, proboscis monkey, Sumatran elephant, tarsier

Plants include the Titan Arum (large flowering plant), rainbow gum eucalyptus tree, melati (a small shrub or vine with fragrant flowers), Anggrek bulan (one of the three national flowers of Indonesia, also known as the moon orchid), Nipa palm, Common swamp pitcher plant, teak tree

Questioning: Responses may include questions about going to school when living on a river, how to bathe/shower, what happens when it floods, and how they do their cooking.

Analysing: It is easy for everyone to access. Since they live on the river, many families have boats for transport. They can buy and sell produce when it is fresh and they do not have to carry big loads.

Communicating:

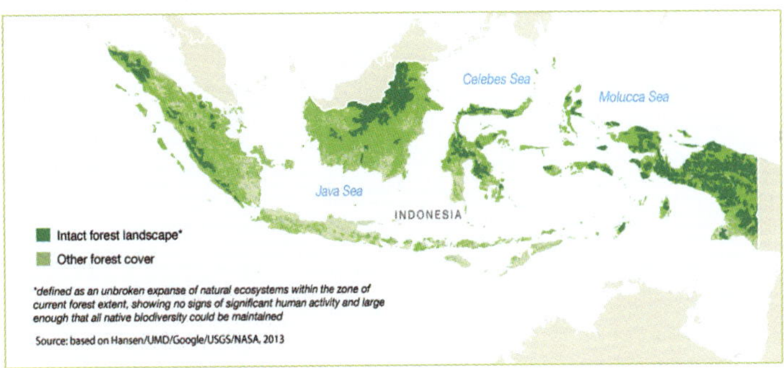

Cartographer: GRID-Arendal, https://www.grida.no/resources/6957

Answers

Note: Answers are not supplied to open-ended questions, where students' responses will vary.

ANSWERS

Evaluating and reflecting: Sample response:

The people live a subsistence lifestyle – living off what the land and river provide. They eat what they can catch and harvest. They do not have facilities such as factories and warehouses for collecting, storing and manufacturing lots of produce, so they make very little money. They spend all their time fishing and farming so they do not have the skills to get jobs in the city and earn more money.

16 Demographic Differences

Research:

Developed nations	Developing nations
Higher levels of population education Increasingly urban and less rural Higher per capita incomes Population levels and population growth are more stable	Low levels of household savings Lower levels of productivity and capital (machinery) Populations tend to be younger on average Higher tariffs (taxes) More people working in rural areas and in agriculture

Questioning: sample responses:

Overall, which gender has the longer life expectancy? What is the average age difference between the life expectancies of men and women in these countries? Which country has the lowest life expectancy? Which country has the highest overall life expectancy? In which country would you be likely to live longer?

Analysing: People with low incomes cannot always afford to go to the doctor and pay for medicines when they are sick. Poorer people also find it more difficult to buy fresh food such as fruit and vegetables to keep them healthy. People on lower incomes have less access to education and information about healthy living options.

Evaluating and reflecting: The government would need to channel more funds into early childhood education. They would need to build more schools and plan for more hospital beds in that area. They would also need to think about recreational facilities, parks and playgrounds in the near future. This would be to cater for an increasing number of younger people growing up in that area over the next 5–10 years.

17 Indigenous Peoples and the UNDRIP

Research: The Akha are one of the smallest and least-developed hill tribe groups in South-East Asia. They live in northern Laos, western Burma, northern Thailand, northern Vietnam and parts of southern China. They are made up of several ethnic sub-groups, but they do not mix among villages and their languages differ quite a lot. They believe that their ancestors provide them with blessings and give them good health, fat animals and good harvests. Some villages have a shaman, or a man of the spirits. They believe that some events such as a tree falling down near a sacred site or a dog climbing on the roof of a house are unlucky.

Questioning:

self-determination: the process by which a person or group of people control their own lives

cultural identity: the feeling of belonging to a particular group. It is related to nationality, ethnicity, religion, social class and generation.

informed consent: giving permission with full knowledge of the possible consequences

declaration: a formal statement of announcement

Proposal: a formal plan or suggestion for others to consider

Analysing: The viewpoints of indigenous people around the world would be similar because many have had explorers from other countries bring disease into their land. They have also lost land to explorers claiming countries for their own leaders, far away. Indigenous peoples would say that their land had not been 'discovered' by these explorers because they had been living there for thousands of years and know the land very well.

18 Fukushima

Research: The recovery and clean-up of the environment is ongoing. Closing down the plant is estimated to cost billions of dollars and take up to 40 years. There is still some contaminated water flowing into the ocean, but the soil contamination was not serious. The topsoil was removed and disposed of, so some farming in nearby areas has resumed. Authorities have been cementing the seabed near the plant to prevent the spread of radioactive material into the sea.

Questioning: Sample questions:

How scared were they? What action did they take immediately? What happened to their own house? Have they experienced an earthquake or tsunami before? What happened immediately after the disaster? How did first responders help them? Have they returned to Fukushima or do they want to?

Analysing: The images help the reader to understand the scale and size of the disaster. Seeing the wall of water tossing around cars gives a sense of awe and an understanding of the incredible power of the tsunami wave. The image of the protester shows the fear that members of the public had about nuclear power and the dangers that come with its use. It helps you to understand that this event did not just impact Japan, but the whole world.

19 Connecting with Asia

Research: Australia imports around $200 billion worth of products from Asia each year: cars 9.1%, refined petroleum 7.2%, broadcasting equipment 3.6%, computers 3.2%.

Australia exports around $245 billion worth of produce to Asia each year: iron ore 20%, natural gas 8.3%, coal 19%, gold 12%, wheat 2%, wool 1.1%.

Questioning: Sample responses:

Will I be able to make friends? Is there cheap accommodation near the university? Can I work part time while I study? How much will the course cost? How easy is it to travel in Australia? What is the weather like?

Analysing: Many Asian students stay because they can get a job in their chosen field of study. They make friends here and form relationships with people. They enjoy the lifestyle in Australia.

TARGETING HASS 6 © PASCAL PRESS ISBN: 9781925726077

Answers

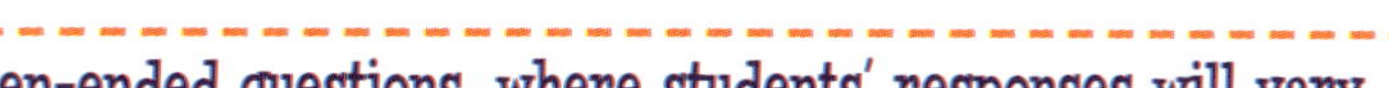

Note: Answers are not supplied to open-ended questions, where students' responses will vary.

Evaluating and reflecting: sample response includes the human need to belong and be part of a group. That we can learn more by sharing our cultures and that can help to break down barriers such as racism and discrimination. We all serve to gain when we can learn new ways to cook, new ideas for solving problems etc.

20 Australia's Aid Program

Research:

Cook Islands: $3.7 million in 2019–20, spent on water monitoring for clean water and sanitation, a minibus to transport disabled people, and supporting education programs

Fiji: $58.8 million in 2019–20, spent on health and education programs and supporting disaster relief after cyclones

Vanuatu: $66.2 million in 2019–20, spent on disaster relief after tropical cyclone damage, improving education and health programs, building infrastructure

Papua New Guinea: $607.5 million in 2019–20, spent on developing a network of national priority roads, the sustainability of the Kokoda Track, improving productivity in the public sector, and training and education programs

Questioning: Sample response:

What was Australia's response to Cyclone Winston? What types of aid does Australia provide to other nations? Why does Australian aid support infrastructure in Papua New Guinea? Why does Australia spend most of its aid budget in the Pacific and South-East Asia?

Analysing: The graph tells you which regions Australian aid money is allocated to, and how much was allocated in the 2016–17 financial year. It is an estimated amount, so is not what was actually spent. The money is allocated in this way to assist countries who are Australia's close neighbours and who are among the poorer countries in the region.

21 Australia's Global Connections

Questioning: Sample responses:

Why do countries establish sister city connections? How much money does Australia give to developing countries each year? Would Australia still participate in the Commonwealth Games if we became a republic? Why do we allow students from other countries to study in Australia? What are some examples of our tourism and environmental connections with other countries? What is it like to live in Australia's Antarctic base?

Analysing: Sample response:

Advantages	Disadvantages
Student exchange programs Sister zoo relationships between Taronga Zoo and Nagoya Higashiyama Zoo Visits by sporting teams and cultural groups Enhances trade and business opportunities	The financial costs Not enough staff on councils to make the program work properly Language barriers

22 First Peoples

Research:

The **Shuar** live in the upper Amazon region of Ecuador and Peru. They live in the tropical rainforests, lowlands and savannas. They are the second largest group of Indigenous people in Ecuador. There is conflict between the Shuar and gold and copper mining companies in the Amazon. They first met Europeans in the 16th century. They were a semi-nomadic people. Men would weave clothing and hunt. Women would garden and tend the crops in the fields.

The **Yanomami** are a group of around 35 000 people who live in villages in the rainforests and mountains on the border between Venezuela and Brazil, on the Orinoco River and the northern reaches of the Amazon River. They are quite isolated. They use slash-and-burn methods of agriculture and live in small, scattered semi-permanent villages. They hunt monkeys, deer and armadillos and grow tubers, corn and other vegetables.

The word **Kogui** means 'jaguar'. These Indigenous people live in northern Columbia. Modern Kogui try to live a traditional lifestyle. They are the descendants of the Tayrona culture, which built stone structures and pathways through the Sierra Nevada mountains. The Spanish did not conquer the Kogui. They are not nomadic. They have cultivated their fields and farms for more than 1000 years.

Questioning: Sample response:

What special rituals do they have? How do they mark time? What special roles do the men and women have? Is there an initiation ceremony for children when they become adults? What is their diet like? What is their housing like?

Analysing: It tells us about their traditional clothing and the types of weapons they used when hunting. It also shows us the type of housing that they lived in and what it was made from.

23 Mark's Diary

Research: The Australian Volunteers program matches skilled Australians with partner organisations in the Indo-Pacific region, with a focus on developing countries. The volunteers work with them to help them achieve their development goals. The group operates in countries like Cambodia, Papua New Guinea, the Philippines, Solomon Islands, Mongolia, Nepal, Tonga, Vanuatu, Fiji and Indonesia. The group pays for return airfares and visas, as well as accommodation costs. Volunteers have to apply and attend an interview to see if they are suitable.

Questioning: Sample responses:

Why did you choose to go to Laos? Was it what you were expecting? What did you learn from your work there? What was the most challenging part of the work? Would you recommend that other people follow in your footsteps?

Analysing: The photographs help show the way of life for the people of Laos and emphasise the fact that Laos is a developing nation. They show the benefits of the Volunteer program and how the people are in need of skilled workers to help them complete important infrastructure, health and education programs.

Answers

Note: Answers are not supplied to open-ended questions, where students' responses will vary.

ANSWERS

24 Cultural Diversity

Research: Angkor Wat is a Buddhist temple complex in Cambodia. It is the largest religious monument in the world. It was originally built as a Hindu temple for the Khmer empire, dedicated to the god Vishnu. It was transformed into a Buddhist temple at the end of the 12th century. UNESCO (United Nations Educational, Scientific and Cultural Organisation) has set up a program to protect this symbolic site. The site is made up of dozens of temples, water basins, canals and communication routes.

Questioning: Sample responses:

Arabic – What is Egypt's official language?

Pchum Ben – What is the Cambodian name of the celebration known as Ancestor's Day?

Tanoura – What is the name of a traditional Egyptian dance?

Khmer – What is the official language of Cambodia?

Ramadan – What is the month-long festival in which Muslims fast?

Analysing: The cultures of Cambodia and Egypt are similar in that one of the other languages spoken in the country is English. They also have significant traditions and celebrations which involve their religious beliefs. These traditions in both countries involve dances and the wearing of brightly coloured traditional costumes.

25 Australians as Peacekeepers

Research: Australian-led peacekeeping missions have been 1. Kashmir (1950–1966), 2. Cambodia (1992–1993), 3. Sinai (1994–1997), 4. Iraq (1997–1999), 5. East Timor (1999–2000), 6. UN truce supervision (1998–2000).

Peacekeepers have generally been unarmed military observers. They make sure that neither side breaks any ceasefire agreements. They report on what they see so that both sides are held accountable for their actions. Police also serve in peacekeeping forces. They are there to maintain public order during times of increased tension, such as national elections.

Questioning: Sample responses:

What is it like being a peacekeeper? Is it hard living away from your family? How dangerous is it? What is the most rewarding part of your work? What is the most difficult part of your role? Why do you enjoy being a peacekeeper?

Analysing: The website itself is a secondary source. But on the site there are some primary sources, such as photographs of soldiers and peacekeepers in action, works of art representing battlefields, portraits of servicemen and servicewomen, and diary entries written by servicewomen and servicemen.

26 Australia's Political System

Research: The Westminster System is the democratic parliamentary system used in the United Kingdom. It has two houses of parliament (lower and upper). The leader of the government, the prime minister, comes from the lower house of elected representatives. The role of the governor-general is to represent the Queen. They give royal assent (approval) to the Acts passed by parliament so that they can become laws.

Questioning:

1. The leader of a local council is called the mayor.
2. The state legislative council is responsible for monitoring and reviewing laws, legislating and gathering information.
3. The role of the governor is to represent the monarch (king or queen).
4. A cabinet is a group of parliamentarians made up of the prime minister and senior ministers.
5. Senators represent their state or territory.

Analysing:

Similarities	Differences
Most have two houses (lower and upper) The leader of the government comes from the lower house They have a cabinet made up of senior ministers They have a representative of the monarch	Queensland and the Northern Territory only have one house Federal has a governor-general, states have a governor State parliament approves laws made by local councils State and federal governments are responsible for different areas (state: education, health, transport; federal: trade, defence, taxation)

27 Australia's Legal System

Research: The idea for a law (bill) is written up by parliamentary officers. It is presented (read) to the lower house. Then it is debated, and some changes may be made. It is then voted on by the lower house in the third reading. If approved, it is read in the senate. Then it is read a second time and debated. It goes to a parliamentary committee to examine the bill and what it intends to do. Changes may be made. It is then voted on by the senate in the third reading. If approved, it goes to the governor-general for royal assent. It then becomes a law.

Questioning:

1. The Family Court hears matters relating to parental disputes, child support and custody.
2. The most powerful court in Australia is the High Court.
3. Three to five judges sit on the Supreme Court.
4. No, the District Court would not hear a case about the theft of a scooter. This would be a matter for local courts and magistrates because it is a lesser offence of burglary.

TARGETING HASS 6 © PASCAL PRESS ISBN: 9781925726077

Answers

Note: Answers are not supplied to open-ended questions, where students' responses will vary.

Analysing: Federal courts use judges because they are dealing with complex cases and lots of information that an ordinary person could not be expected to know about or understand all of. Also, federal court cases take a very long time and it is not reasonable to ask a person to stop their own work for such a long time to listen to a case in the Federal Court.

Communicating:

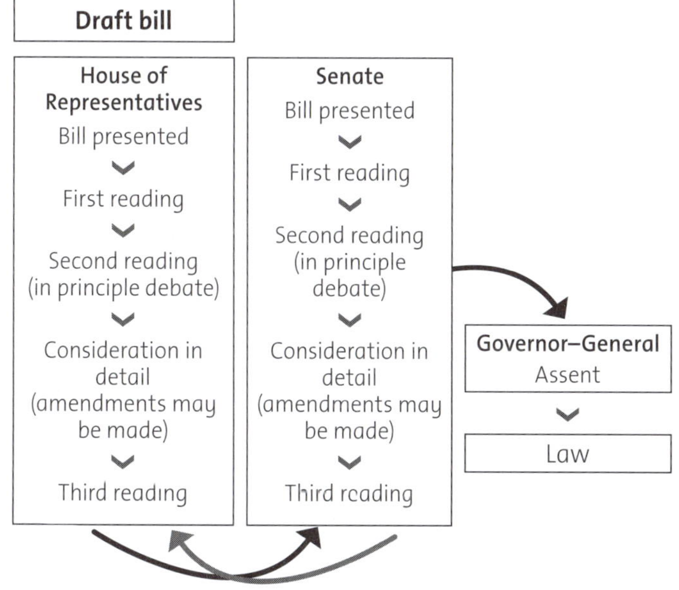

28 Being an Australian Citizen

Research: Being a dual citizen means that you are considered to be a citizen of more than one country. If you are a dual citizen (one of which is Australian citizenship), you may not receive full assistance from Australian consulates if you need help elsewhere in the world. You should also use your Australian passport, not a passport from your own country of citizenship, when travelling into and out of Australia.

Questioning: Sample responses:

How long do you need to live in Australia before you can apply for citizenship? Where were most Australians born, other than Australia? What percentage of Australians were born in India? How did Prime Minister Keating change the citizenship oath?

Analysing: Over the years, the citizenship has changed from an oath to a pledge. It started by swearing allegiance to the King, then to the Queen, and then to the people of Australia. The current pledge also has an option to remove the words [under God] for those who have different religious beliefs.

Communicating:

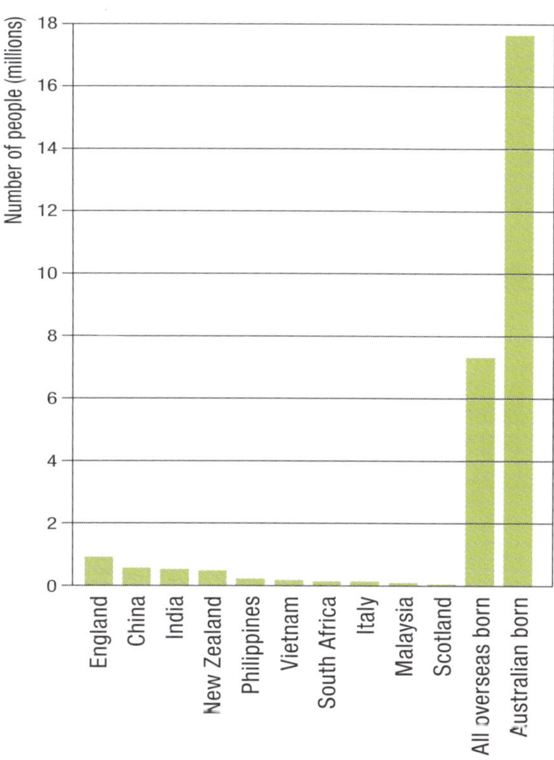

29 Trade-offs and Costs

Research:

In 2018, Cadbury changed the flavours and wrapping of its Roses chocolates. Dropping some of the traditional flavours received a lot of public backlash. There was a petition to bring back the classics, such as the strawberry crème and hazelnut whirl. Some of the newly introduced varieties couldn't be eaten by people with food intolerances such as coeliac disease.

In 2016, Arnotts changed the recipe for their Shapes brand of savoury biscuit. There was a change.org petition with more than 10 000 signatures that demanded the company revert to the original recipe. The company later announced that it would keep both the new product line as well as reintroduce the old favourites, such as original barbeque and chicken crimpy.

Questioning:

Arnotts changed their product after they conducted several trials of new flavours when their customers asked for bigger and stronger flavours.

Cadbury also said that they changed the original Roses based on customer feedback.

The trade-off was that the companies spent time and money creating new recipes instead of keeping the original. The opportunity cost of the decision was the number of people who would no longer buy the product and who would buy a different brand of product instead.

Answers

Note: Answers are not supplied to open-ended questions, where students' responses will vary.

Analysing:

Arnotts

Positive	Minus	Interesting
Gained media coverage of their brand Got people talking about the product Now have two types of Shapes (new and original)	Upset customers Time wasted on developing new recipes and dealing with social media complaints	They kept both the old and the new products Now, how much of each do they sell?

Cadbury

Positive	Minus	Interesting
Some people liked the taste of the new flavours Gained media coverage of their brand Got people talking about the product	People with food issues can't eat them Lots of people complained People like the familiar People don't like change	Are they selling as much or more of the new version as they did of the old version?

30 Making a Profit

Research: Care Australia is a humanitarian aid organisation that fights global poverty. It has a focus on helping women and girls.

The Australian Red Cross is a humanitarian aid and community service charity. They have been operating for 106 years. They help with bushfire recovery, mental health and homelessness, and migrants in transition.

The Fred Hollows Foundation was founded in 1992. It focuses on treating and preventing blindness and other vision problems. It operates in Australia, focusing on Indigenous communities, the Pacific, South-East Asia and Africa.

The Koorie Heritage Trust is an Indigenous Australian not-for-profit cultural organisation. It is based in Melbourne. It supports, promotes and celebrates the culture of the Indigenous peoples of south-east Australia. They hold workshops, walking tours and training programs.

Questioning:

retained: to keep possession of

tax: a compulsory financial charge

utilities: organisations that maintain the infrastructure for public services such as water and electricity

deductible: an expense which is subtracted from gross income

revenue: the income generated from operating a business

expense: the cost of purchasing something or operating a business

Analysing: Not-for-profit organisations may find it difficult to get people to volunteer to help. People might be willing to donate money as a one-off event, but not in an ongoing or regular way. This means that not-for-profit companies will need to spend more of their money paying people to do the work and so they can do less with the money that they have.

31 Printer Cartridge Recycling

Research: The cartridges 4 planet ark program has operated since 2003. It has recycled more than 45 million printer cartridges. It turns them into road asphalt additive, garden beds and pens. The program works with Brother, Canon, Epson, and HP to help make them be more responsible for their products at the end of their useful life. There are public drop off boxes or people can request one for their school or business.

Questioning: sample questions may ask – How often do you recycle? Do you recycle your printer ink cartridges? What types of products do you recycle? How do you recycle these products? Do you buy products made of recycled materials? What are they?

Analysing: The image of computer waste shows the different types of materials that can be recycled. It also makes the reader feel that it is a mess that needs to be fixed or cleaned up. It was included to get an emotional reaction from the reader: either sad about the waste, angry about what people throw out and the cost of not recycling our waste.

32 Global Financial Crisis

Research: The federal government gave $750 cash to pensioners and welfare recipients. Businesses were able to deduct large purchases up to $150 000 off their tax. They allowed workers to have access to their superannuation funds early and withdraw up to $20 000 if they suffered financial hardship due to the virus. The income support for the JobSeeker payment was doubled. They created a Jobkeeper payment of $1500 a fortnight. It was paid to employers who were then asked to pay it on to their workers to keep them in a job. People receiving a parenting payment were given an additional $550 a fortnight for six months.

Questioning: Sample responses:

What is a recession? When was the GFC? What caused the 2020 recession? How was Australia impacted by the GFC? Why did the government introduce the Jobkeeper payment?

Analysing: The trade-off was not being able to spend that same amount of money on other things, such as international aid, Medicare payments to doctors or defence.

Communicating:

Covid
* world wide
* health and financial crisis
* borders closed

GFC
* credit crunch
* started in USA
* Australia not badly affected

BOTH
* government stimulus
* long-lasting impacts

TARGETING HASS 6 © PASCAL PRESS ISBN: 9781925726077

Answers

Note: Answers are not supplied to open-ended questions, where students' responses will vary.

History Assessment 1

1. 1901

2. Students need two out of these four answers.

Name: Henry Parkes	Name: Charles Kingston
Role in Federation: Gave the Tenterfield Oration to call the public to support federation.	Role in Federation: Helped draft the first Australian constitution and travelled to London to present it to the British parliament.

Name: Edmund Barton	Name: Alfred Deakin
Role in Federation: Gave speeches supporting federation all over NSW. Was a delegate at the Sydney constitutional convention.	Role in Federation: Attended federal conferences and conventions. Travelled around the country giving speeches in support of federation.

3.

Reasons to support federation	Reasons to oppose federation
Federation would remove taxes and allow free trade. There would be one army. There would be one postal service and one rail gauge across the country. Australians could make their own laws and rules, instead of following British laws. People born here wanted to be recognised as Australian, not British.	Rich colonies did not want to share their wealth with poorer colonies. Poorer colonies liked being able to tax goods from the other colonies. Remote and poor colonies thought the larger colonies would not listen to them in a federation. It was too hard. There were lots of decisons. Where would the capital city be?

Australia's involvement in World War I	We would not have been involved as a nation. There would be no Anzacs. We would not have been led by our own officers. States may have fought, but as separate colonies.
Establishing Aboriginal citizenship rights	Without a federated Australia, there would be no national citizenship rights for Aboriginal people. The states would make their own laws and they may not be consistent across the country. Aboriginal people may still not be included in the census.
The development of 6 states and 2 territories	Without federation, the Northern Territory may still not exist. It might still be part of South Australia. The ACT would not exist either.

History Assessment 2

The Australian Constitution: The document mentioned Aboriginal peoples only twice. Torres Strait Islander peoples were not mentioned at all. Section 51 stated that the federal government could not make laws for Aboriginal people. Section 127 stated that they were not to be counted in the census. These two sections made life very difficult for Aboriginal peoples. They did not have the same rights and freedoms as other Australians.

Aboriginal missions: In the nineteenth century, religious organisations set up areas around Australia known as missions. The goals of these organisations were to protect, house and convert Aboriginal people to Christianity. They were run by missionaries and discouraged the expression of Aboriginal spirituality and culture. Over time, many missions were taken over by the government and were replaced with reserves or stations. On these stations, managers controlled who could enter and leave, and also where Aboriginal people could live and go to school.

State Aboriginal Protectors: The states created a position called Chief Protector of Aborigines. In Queensland it was called Protector of Aboriginals. The protector had to look after the rights of Aboriginal people, guard their property and protect them from cruelty, oppression and injustice. Although the role was supposed to protect Aboriginal people, particularly in remote areas, it ended up being a severe form of social control. The protector had power over who Aboriginal people married, where they lived and how they managed their financial affairs. Aboriginal peoples had no control or power over their own lives and choices.

The 1967 referendum: The referendum asked the Australian people if they approved of the proposed law that would allow Aboriginal people to be counted in the census. The change to the constitution in the referendum aimed to give the federal parliament power to make laws with respect to Aboriginal people. It deleted part of section 51 of the constitution and repealed section 127. The result was a strong YES vote with an overall majority of 91 per cent.

History Assessment 3

1. The image is a primary source.
2. It was created in 1947.
3. The photo was taken to show the living conditions in the European refugee camps and to gain the audience's sympathy for their situation.
4. It tells us about how they lived, the types of housing they had, that there were children living in the camps and that they had very poor facilities for cleaning and washing.
5. Yes, the source is reliable. It is a photograph of a real event, not a painted image or a cartoon, relying on the artist's impression and skill.
6. Marija's push factors were the war and the fact that life in Latvia was very dangerous. The Russians were treating people very badly.
7. Marija's pull factor was that Australia had agreed to take them – they had permission.
8. A refugee is someone who flees their home due to war, poverty or persecution.

Answers

Note: Answers are not supplied to open-ended questions, where students' responses will vary.

9. Marija feels frightened because she doesn't know anything about Australia and she doesn't speak the language.

Geography Assessment 1

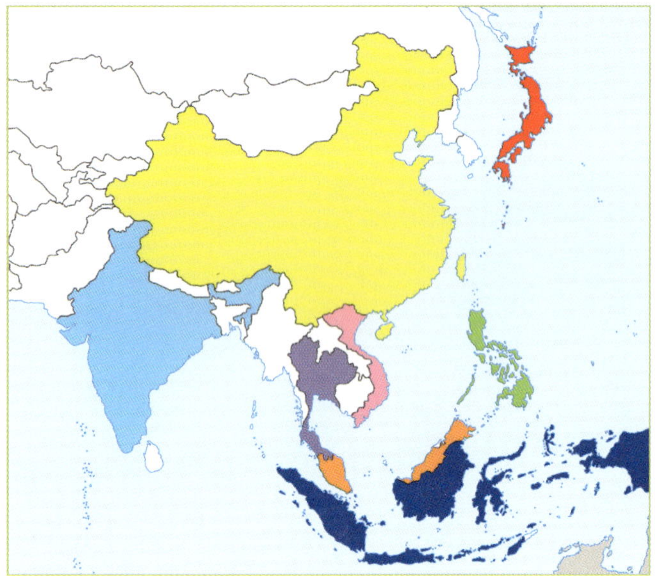

	Country	Population	Capital city
	Japan	126.8 million	Tokyo
	Thailand	69.04 million	Bangkok
	India	1.339 billion	New Delhi
	China	1.386 billion	Beijing
	Philippines	104.9 million	Manila
	Malaysia	31.62 million	Kuala Lumpur
	Vietnam	95.54 million	Hanoi
	Indonesia	264 million	Jakarta

Geography Assessment 2

1. The Iban people live in Asia.
2. The climatic zone is tropical.
3. They would build along rivers and creeks to give themselves a fresh water supply. There would also be lots of fish to catch and they could hunt the animals that came to drink.
4. They would have access to trees from the forest plus any that washed down the river. There would be lots of palms and ferns to use as thatching.
5. The weather conditions would be hot and humid. There would be monsoonal rains.
6. The climate would influence the types of clothes they wear and the types of fruits and vegetables that grow. It would also influence how they build their houses – the one in the picture has raised stilts to protect them from river flooding. The environment would affect the way they live through the materials they have access to for clothing, shelters and tools, and the natural resources they had for making medicines. Some things in the environment would bring danger, such as spiders, snakes and flooding.

Geography Assessment 3

1 T: A developed country has strong industries such as manufacturing and banking.

F: Developing nations have access to good health care and education.

F: Developing nations have populations that are low and do not grow quickly.

T: The average income in developed nations is high.

T: Developing nations have high population densities.

F: Developing nations have higher life expectancies than developed nations.

T: Developed nations have low unemployment.

F: Stable government is more common in developing nations.

2. Sample response:

Trade: Thailand is Australia's 6th largest two-way trading partner. A lot of Thai students come to study in Australia. We import cars, automotive components, electronics and processed food and beverages from Thailand. Thailand also exports large amounts of rice, palm oil, rubber and seafood. Australia's main exports to Thailand are crude petroleum, gold, aluminium and coal.

Tourism: Approximately 800 000 Australians visit Thailand each year. More than 72 000 Thai people live in Australia. The most popular Thai destinations are Bangkok and Phuket.

Defence: Australia and Thailand have a defence cooperation program. It costs about $5.7 million each year. The two armed services have worked side by side in United Nations operations in Cambodia, Somalia and East Timor.

Culture: There are many cultural exchanges between the two countries. In 2017, a troupe of Thai Khon dancers held an exhibition at the Sydney Opera House. There are also many interschool exchange programs. Thai and Australia school teachers link up for a two-week professional development program under the Australia-Thailand BRIDGE project.

Civics and Citizenship Assessment

1. The role of the senate is to represent the states in Federal Parliament.
2. The government needs a majority in the lower house.
3. Australia is known as a parliamentary democracy.
4. The leader of a state government is called the Premier.
5. The role of a local council is to manage public property, services and activities and to make by-laws for its district.
6. The federal government is responsible for defence, trade, border protection, customs, foreign affairs, immigration, the environment and taxation.
7. The state government is responsible for health, education, transport, state police, agriculture, fishing, prisons, emergency services, sport and recreation.
8. Australia's head of state is the Monarch (King or Queen).
9. 6 - Once approved by the lower house, the bill goes to the Senate for the first reading.
   2 - The bill is introduced into the lower house (1st reading).
   8 - The bill then goes to a Legislation Committee Public Hearing.
   7 - The Senate considers the bill at the second reading and debates the matter.

Answers

Note: Answers are not supplied to open-ended questions, where students' responses will vary.

1 - A bill (proposed law) is drafted by lawyers in the Office of Parliamentary Counsel.
4 - The lower house considers the bill for changes and amendments.
3 - The bill is read for a second time and debated by the parliament.
9 - After the third reading the bill is voted on by the Senate.
10 - The bill finally passes to the Governor-General for Royal Assent.
5 - At the third reading the members of the House of Representatives vote on the bill.

10. Being an Australian citizen means that you are entitled to the protection of the Australian government and you can exercise your rights and responsibilities as an Australian.

11. The responsibilities of an Australian citizen are: to obey the law, defend Australia if needed, and vote in federal and state elections.

12. Australian citizens are entitled to: apply for an Australian passport, live in Australia, seek assistance from Australian diplomatic representatives while overseas, leave and return to Australia without applying for a visa.

Business and Economics Assessment

1. The trade-off Mikel made was choosing the handlebar grips instead of the lime and berry slushie.
2. The opportunity cost was the loss of enjoyment of the lime and berry slushie.
3. Answers will vary
4.

Human Resources	Capital Resources	Natural Resources
worker training	hammer	water
leadership	computer	iron
recruitment services	3D printer	coal
safety officer	packing boxes	natural gas
	filing cabinet	sand
	welding machine	wood
	crane	raw wool

5. Scarcity: the limited availability of a good, product or raw material, or lack of money to purchase an item.

Profit: the money remaining from income or revenue after all expenses have been paid.

Business: an organisation that sells goods or services to make a profit.

6. Sample answers:

Newspapers, magazines, cards, stationery, wages, electricity, rent, insurance, signage, advertising, paper delivery.

Targeting HASS
Year 6

ISBN: 978 1 925726 07 7
Reprinted 2022, 2025

Published by Pascal Press
PO Box 250
Glebe NSW 2037
contact@pascalpress.com.au

Author: Merryn Whitfield
Publisher: Lynn Dickinson
Typesetter: Ruth Schultz
Editor: Ruth Schultz
Designer: Janice Bowles

Printed by Vivar Printing/Green Giant Press